WATCHING YOUTUBE: EXTRAORDINARY VIDEOS BY ORDINARY PEOPLE

An anonymous musician plays Pachelbel's *Canon* on the electric guitar in a clip that has been viewed over sixty million times. *Dramatic Gopher* is viewed over sixteen million times, as is a severely inebriated David Hasselhoff attempting to eat a hamburger. Over eight hundred variations, parodies, and parodies-of-parodies are uploaded of Beyoncé Knowles's *Single Ladies* dance. Tay Zonday sings *Chocolate Rain* in a video viewed almost forty million times and scores himself a record deal. Obama girl enters the political arena with contributions such as '*I Got a Crush ... On Obama*' and gets coverage in mainstream news networks.

In *Watching YouTube*, Michael Strangelove provides a broad overview of the world of amateur online videos and the people who make them. Dr Strangelove, the Governor General Literary Award–nominated author who *Wired* Magazine called a 'guru of Internet advertising,' describes how online digital video is both similar to and different from traditional home-moviemaking and argues that we are moving into a post-television era characterized by mass participation.

Strangelove draws from the critical literature of television, film, cultural, and media studies to help define an entirely new field of research. Online practices of representation, confessional video diaries, gendered uses of amateur video, and debates over elections, religion, and armed conflicts make up the bulk of this groundbreaking study, which is supplemented by an online blog at strangelove.com/blog. An innovative and timely study, *Watching YouTube* raises questions about the future of cultural memory, identity, politics, warfare, and family life when everyday representational practices are altered by four billion cameras in the hands of ordinary people.

(Digital Futures)

MICHAEL STRANGELOVE is an adjunct professor in the Department of Communication at the University of Ottawa.

MICHAEL STRANGELOVE

Watching YouTube

Extraordinary Videos by Ordinary People

UNIVERSITY OF TORONTO PRESS
Toronto Buffalo London

© University of Toronto Press Incorporated 2010
Toronto Buffalo London
www.utppublishing.com
Printed in Canada

ISBN 978-1-4426-4145-7 (cloth)
ISBN 978-1-4426-1067-5 (paper)

Printed on acid-free, 100% post-consumer recycled paper with
vegetable-based inks.

Digital Futures Series

Library and Archives Canada Cataloguing in Publication

Strangelove, Michael, 1962–
    Watching YouTube : extraordinary videos by ordinary people /
    Michael Strangelove.

(Digital futures)
Includes bibliographical references and index.
ISBN 978-1-4426-4145-7 (bound)   ISBN 978-1-4426-1067-5 (pbk.)

1. YouTube (Electronic resource).   2. Internet videos – Social aspects.
3. Online social networks – Social aspects.   I. Title.   II. Series: Digital
futures.

HM851.S774 2010   303.48'33   C2010-900709-3

University of Toronto Press acknowledges the financial assistance to its
publishing program of the Canada Council for the Arts and the Ontario
Arts Council.

 Canada Council   Conseil des Arts 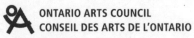 ONTARIO ARTS COUNCIL
for the Arts   du Canada   CONSEIL DES ARTS DE L'ONTARIO

University of Toronto Press acknowledges the financial support for its
publishing activities of the Government of Canada through the Book
Publishing Industry Development Program (BPIDP).

For Anne, the one I like to watch the most

Contents

List of Illustrations

Acknowledgments

In keeping with the subject matter of this book, see the YouTube video *Watching YouTube: Acknowledgments* for a visual tribute to those who made these words about moving images possible.

WATCHING YOUTUBE: EXTRAORDINARY VIDEOS BY
ORDINARY PEOPLE

Introduction

I like to watch. I confess – I enjoy watching laughing babies, home-made cartoons, dancing girls, ranting POOGS (pissed off old guys), the antics of other people's pets, the stupidity of young male jackasses, politicians, and celebrities caught being human, clever student art projects, compelling amateur documentaries, the Têtes à claques, and real-life action from the front lines of the war du jour. Along with hundreds of millions of other people, I like to watch videos on YouTube and other Internet sites. Frankly, you would have to be dead inside not to find something emotionally or intellectually compelling on YouTube. After all, it is you, it is me, it is our neighbours, our families, our friends (and, all too often, our darn kids) who can be seen on YouTube. It is ever more of the world brought to our computer screens by amateurs and everyday people armed with digital cameras, camcorders, webcams, and cellular phones.

Watching YouTube is an investigation into the world of ordinary people and their extraordinary online videos. Whereas the vast majority of television and film studies in the last century examined commercial products of the entertainment industry, this analysis focuses on videos made by ordinary people: amateurs working outside the institutional structures of the television and movie industry. Internet users watch more than 1 billion video clips on YouTube every day, and much of that viewing time is spent watching amateur videos.

We know that over fifty years of watching commercially produced television programs and films had a dramatic impact on life in the last century. How is watching YouTube different from watching television and cinema? What happens when cultural production shifts from being dominated by commercial industries to amateurs armed with cameras and Internet connections? What can YouTube tell us about the future of a society where the audience spends a great deal of time consuming non-commercial forms of entertainment, news, and art? In the game of politics a candidate's every gaffe now is instantly relayed to voters by amateur videographers eager to embarrass the opposition. America's culture war is heating up as all sides arm themselves with video cameras, irreconcilable opinions, and the tactics of Michael Moore. Wired teenagers – tomorrow's leaders – are busy creating YouTube videos of each other dancing in their underwear, drunk, or behaving like skanks and jackasses. Our most embarrassing moment may wind up as the most viewed video of the day.

Watching YouTube cannot capture all of the 'Tube' but it does provide an overview of YouTube's social features. Beginning with a brief survey of the golden age of home movies, *Watching YouTube* explores how online digital video is different from the film-based domestic moviemaking of the mid-twentieth century. Along with mom and dad, children and teenagers are now making home movies and putting them on the Internet. Here we explore these new media practices of representing home and family life. Also, video diaries, the very epitome of YouTube as a social phenomenon, and women's self-representational video practices will provide further insight into YouTube as a domain of self-expression, community, and public confession.

YouTube is not merely an archive of moving images. It is much more than a fast-growing collection of millions of home-made videos. It is an intense emotional experience. YouTube is a social space. This virtual community reflects the cultural politics of the present times and thus is rife with both cooperation *and* conflict. Herein we will also explore the cultural wars of YouTube. Within the YouTube community we can observe racist, sexist, homophobic, and verbally violent debates over elections, religion, and armed conflicts. We once participated in conflict as audiences of nighttime newscasts, documentaries, television dramas, and movies. Now the YouTube generation actively represents democracy, faith, and war through their amateur video practices. YouTube is not merely a new window on the frontlines of regional and global conflicts. It has become a battlefield, a contested ground where amateur videographers try to influence how events are represented and interpreted.

Last, *Watching YouTube* will consider the possibility that we are moving into a post-television era. We are not about to see a giant as huge as television disappear from the media landscape. Nonetheless, amateur video itself has become another giant among us, demanding our attention, fragmenting audiences, worrying advertisers, troubling television executives, and eroding the monopolization of representation by media corporations.

Sitting in a coffee shop one spring day in 2008, I overheard a young woman suggest that her friend 'YouTube it.' What she meant was clear enough – she wanted her friend to see what she herself had seen on a YouTube video. As Google did earlier in this still new century, YouTube had entered the lexicon of the wired generation and was transformed from a noun to a verb. What I had witnessed was more than the bastardization of the English language that causes so much concern among educators and parents.

On the evolution of the use of YouTube from a noun to a verb, in 2006 Lee Gomes noted in the *Wall Street Journal* that YouTube was being used as an adjective, as in 'YouTube election,' and speculated, 'Its final triumph as a verb can't be far behind.'[1] The rise of YouTube as verb, as a part of everyday action, marks a very significant development in communication technology and culture. This is a change that not only will have but already is having a dramatic impact on many aspects of modern life. Although such hyperbole has plagued the study of the Internet from the early 1990s, it is my goal in this book to argue that in this instance, the times are indeed changing.

When television first appeared, it would have been a mistake to proclaim that RCA was going to change the world. RCA and the early television networks were part of a fundamental transformation of industrial society into a broadcast-based media culture that would eventually span the entire globe. The story of broadcast communication – radio, television, and newspapers – is central to any explanation of how life was organized and experienced in the 1900s.[2] Yet it was not any one television network, radio station, or newspaper that defined the unique character of the twentieth century. It was the structural and economic character of the mass media itself that shaped how the baby boomers understood themselves and their world. Content that originated from the commercial media sector promoted the values necessary for the management of consumer habits and the reproduction of capitalism. To suggest otherwise is to claim that the collective efforts of the advertising and public relations industries have been a colossal waste of money.

It would be extraordinary if corporate mass media worked to undermine the values of its sponsors. The simple fact that billions of people spent so much of their lives consuming content from corporations in a largely one-way relationship explains much about the last century. Yet this advertising-driven relationship between content and audiences is changing. The change that I speak of is not solely because of YouTube. It would be the worst sort of hype to claim that it is YouTube itself that heralds a new type of social order. The medium, not the corporation, is the message we need to heed. YouTube is one of the most visible manifestations of a widespread change in how the Internet and a plethora of related digital technologies are being used. At the centre of this change are individuals and their amateur videomaking practices. This change is also part of a long-term transition in the nature of the audience from relatively passive consumers to fully active producers of moving images.

In the 1900s the advertising and public relations industries spent trillions of dollars on behalf of corporate and government clients intent on shaping opinion, politics, and consumer habits. This carefully manufactured visual representation of events, things, and people now must compete with the visual representation of the mundane, uncensored, unsponsored counter-discourse of our 'Tubes, our home-made visual representations. A social order based on controlled representation of events is discovering the uncertain consequences of uncontrolled representation. Television is losing eyeballs to amateur video. Consumer-generated content and amateur cultural production also affect all other aspects of corporate communications. Advertising, news, and public relations – vital to the maintenance of consumer society – are faced with a tsunami of amateur video.

YouTube is a commercial Web site launched in June 2005 and bought by Google in 2006 for $1.65 billion (U.S.). Google makes $16.6 billion (U.S.) in annual revenues from online search advertising and anticipates turning YouTube into a lucrative property for advertising. YouTube makes the majority of its revenue by delivering audiences to advertisers. By summer 2009 YouTube had an estimated annual operating cost of $710 million (U.S.), half of which resulted from bandwidth costs, and had yet to make a profit.[3] Critics claim that Google grossly overpaid for YouTube and the site will never make a reasonable profit. Like any new Internet venture, its future is far from certain.

The majority of amateur videos that YouTube hosts are not sponsored by advertising. In many cases amateur video represents material that is of little interest to advertisers. According to *Forbes* magazine, 'Advertising hasn't taken off on YouTube because advertisers are worried that their brands will appear next to questionable content.'[4] The online audience is not fond of advertising. Advertisements that play prior to the main video cause 70 per cent of the audience to stop watching the video.[5]

Google chief executive, Eric Schmidt, has acknowledged that 'Google hasn't figured out how to make lots of money from the video site.'[6] Indeed, the advertising industry is experiencing considerable difficulties 'monetizing' user-generated content found on amateur video and social networking sites.[7] Internet videographers and amateur content creators are aware of the trend towards integrating advertising into and around their content and are concerned. Schmidt was seen trying to allay such fears when he claimed, 'The goal of the company isn't to monetize everything. The goal of the company is to change the world.'[8] It remains to be seen if Google and YouTube can change the world when they also intend to further entrench the audience in the consumer lifestyle.

YouTube has invited popular members to become partners and add advertising to their videos. This has allowed a few individuals, such as Michael Buckley, to make a living from their YouTube channels. Nonetheless, in 2009 YouTube was selling advertisements against no more than 9 per cent of video views in the U.S.[9] YouTube's advertising revenue was so low that a May 2009 edition of *Time* magazine included YouTube among a list of 'the most colossal tech failures of the last decade.' This judgment that YouTube had 'clearly missed the mark of living up to the potential that its creators expected, and that the public and press were lead to believe was possible' is premature, highly subjective, and wildly inaccurate.[10] It is far too early to start writing YouTube's obituary.

The awkward fit between advertisers' needs and user-generated content will not diminish the importance of amateur video to commercial enterprises such as YouTube. Google is certainly aware that it needs a host of amateurs who are willing to post their videos on YouTube. Without the willing and abundant contributions of amateurs YouTube would find it much more difficult to compete for our attention. The attraction of YouTube is that it contains both commercial content – what we see on television and at the theatre – as well as non-commercial content.

YouTube represents a deepening interrelationship between user-generated content such as amateur video and commercially produced content. As much as YouTube must find a way to deliver audiences to advertisers, it must also continue to provide the audience with an alternative to the usual viewing fare found on television. The irony of YouTube is that, as an advertising-driven commercial enterprise, it demonstrates the strength of the audience's desire for an alternative to commercially produced content.

Amateur video provides an alternative to content that has been created by professionals for the purpose of segmenting audiences into digestible target markets for the advertising industry and their clients. Mass participation in amateur video production suggests that there is something new within media culture. Never before have so many people across the globe spent so much time viewing so many videos made by amateurs. The study of the Internet is riddled with claims about 'new' and 'revolutionary' phenomena, which upon subsequent investigation have been demoted to the more humble categories of 'innovative' and 'significant.' In light of the recognition that 'many of the characteristics we consider unique to the Internet are not new at all,' Chris Atton suggests that we 'take care not to lionize the Internet as wholly new.'[11] Given the excesses of the past, this is sound advice, but media theory risks overlooking seismic shifts in the social landscape if it fails to identify truly new spaces of cultural production and consumption. We are on solid ground when we label mass participation in amateur video as a new mode of production and consumption. Amateur video provides an alternative to commercially driven content produced by professionals labouring within the entertainment and media industries.

Of late, much has been written about user-generated content, as if this is some new trend, but the Internet has always been about content created by its users. The blogging phenomenon that swept through the media system and worried journalists and publishers alike was much ado about user-generated content, as is today's fascination with social networks and amateur video. In 2006 the cover of *Time* magazine, which was fashioned in the image of a YouTube video, declared, 'You' are the person of the year.[12] In so doing, *Time* was paying tribute to the transformation of the mass audience from passive watchers to active producers of compelling (and disgusting, boring, trivial, educational, hilarious, and creative) amateur videos.

It is indicative of the sacred nature of the screen in media culture that *Time* did not declare 'You' (that is, 'us,' everyday individual content producers) as persons of the year when the Internet first rolled over mass media in the heady days of the 'Information Super Highway' or during the explosive appearance of the Web, or in response to the chattering blogging classes. None of these text-based forms of Internet use garnered *Time*'s person of the year award until the unwashed masses could mimic the sacred icon of media culture – the screen. The Internet user, as seen on something-like-television – now that has our collective attention!

The Internet is evolving into a television-like medium but is doing so without the same structural, economic, and power relations that made television a tool of economic and political power. The control of television created certain types of power and sustained particular forms of relationships. This 'TV power' was and is predominantly monopolistic – only certain classes have access to owning and operating the production infrastructure of television. While having a video camera and an Internet account is not the same thing as owning a television station or film studio, this new production and distribution system does give the amateur videographer certain aspects of TV power. When audiences watch and contribute to YouTube they are participating in an ancient form of representational power – the power to tell their own stories.

Watching YouTube surveys the variety of social uses of amateur video that are redistributing TV power. The primary theoretical lens used herein is that of the digital ethnographer.[13] A media ethnographer is one who deeply participates in the community being studied and actively uses the media technology under investigation. I approach online amateur video as a cultural theorist, as an amateur videographer who has been using YouTube as a teaching tool, as an artist who creates experimental online videos, and as a blogger who writes about amateur culture and the 'Tube. Descriptions of field research still privilege the contexts of face-to-face interactions and thus miss how social interaction is taking place online.[14] Media ethnographers typically immerse themselves in the media they study, join the communities enabled by those media, and participate in the productive activities of the community. An ethnographer's text, such as *Watching YouTube*, is said to gain its authority from the writer's personal experience. Yet there are a multitude of online communities and experiences to be had within YouTube. Thus, no one text can authoritatively represent the people, communities, and culture of the 'Tube in their entirety. As Jean Burgess and Joshua Green note in *YouTube: Online Video and Participatory Culture*, 'each study of YouTube gives us a different understanding of what YouTube actually is.'[15]

YouTube represents a transformation in the structure of our media-saturated culture. This transformation is both simple and profound in its consequences. This is the transformation of *who is saying what to whom.* While this assessment of the Internet's function may sound no different from what many wrote in the early 1990s, there nonetheless remains a new element to current patterns of Internet use and that is the element of video. When billions of individuals already spend so much time watching television, mass participation in amateur video adds another highly seductive dimension to media culture (or as some have called it, 'screen culture'). In many ways this previously absent dimension is 'You' – me, that girl sitting next to you on the bus, all those people with digital cameras built into their cellular phones, computers, laptops, iPhones, Blackberries, and teddy bears. You in full motion, ever closer to broadcast quality, and ever more likely to appear on YouTube or some other Web site.

On a daily basis the old media of newspapers and television are bringing to our attention just how many ways the new medium of amateur video is being used. As I was writing this introduction, CNN reported that a young girl named Crystal made a YouTube video that explained how she was raped but, in her opinion, denied justice. BBC News reported that a gang member was arrested in Dade County, Florida, after threatening the police via a YouTube video. A few weeks earlier every newspaper in North America was covering the story of six teenage girls in Florida who had recorded their brutal attack of another girl and posted the video on YouTube. Stories like these put amateur videos in the news on a daily basis.

With over 150 million videos hosted, the uses of YouTube expand with each passing week. Randomly exploring YouTube is like channel surfing through 100 million lives. A British playwright used YouTube to publish a video complaint that provided intimate details of her husband's alleged terrible behaviour. Two supercandidates in the American presidential primaries announced their support for Barack Obama via a YouTube video. Some soldiers used YouTube to document their killing of enemy forces in Iraq. Iraqi insurgents used YouTube to document their killing of American forces. Thousands of people uploaded cute videos of their pets. Kids created videos of schoolyard fights. An epileptic woman posted a video of her seizure. Every day more videos from the YouTube community arrive in our email boxes and over ninety of them are discussed herein – a tiny slice of a vast mass of digital humanity.

There are conflicting reports about exactly how many people watch online video, but there is a general consensus that the number is significant and growing.[16] YouTube claims that twenty hours of video are uploaded to its servers every minute – which suggests that 365,512 videos are uploaded every day.[17] This is the equivalent of Hollywood releasing 114,400 new full-length movies into theatres each week. In fall 2009 YouTube served a total of 1 billion videos a day worldwide (and possibly as high as 80 billion a month).[18] A whopping 79 per cent of YouTube videos are estimated to be user-generated content.[19]

Cisco Systems expects that online video will cause Internet traffic to increase fivefold over the next five years.[20] Video makes up 32 per cent of all global consumer Internet traffic and is expected to exceed 90 per cent by 2013. Digital screens are growing in size and proliferating: 'By 2013 the surface area of the world's digital screens will be nearly 11 billion square feet (1 billion square metres), or the equivalent of 2 billion large-screen TVs. Together, this amount would be more than fifteen times the surface area of Manhattan. If laid end to end, these screens would circle the globe more than 48 times.'[21] Cisco notes that its forecast is considered conservative by most industry analysts. AT&T claims that 40 per cent of its own network is filled with video traffic and expects that figure to rise to 65 per cent over the next five years. Clearly, the Internet is transitioning to a predominantly video-based medium.

In 2006 approximately 70 per cent of YouTube viewers were American.[22] By 2008 only 30 per cent of YouTube viewers were American. In April 2009 more than 152 million Americans viewed online videos 16.8 billion times, including 107 million who watched 6.8 billion YouTube videos in one month (63.5 videos per viewer). By Christmas 2009 over 82 per cent of American Internet users watched some video online and the average viewer watched ten hours worth of videos per month. Hulu, a site that allows viewers to watch only commercial media, attracted less than 3 per cent of the online audience for videos in 2009, while YouTube's mix of commercial and amateur videos commanded over 40 per cent audience share. YouTube's closest competitor, Fox Interactive Media, commanded a mere 3.1 per cent audience share. YouTube videos migrate across the Internet and are seen by the majority of the total global Internet population. Average video length continues to increase and is now over 3.5 minutes.[23] There is almost complete gender parity among YouTube's visitors (only 1 per cent more male than female viewers).[24] Although these statistics are now out of date, they provide a benchmark for the stunning growth rate of online video. YouTube is partnering with movie studios and television networks to provide more long-form content such as movies and television shows; thus, the average length of online videos and the time spent watching them will increase dramatically.

An IBM survey of Australia, Germany, India, Japan, the United Kingdom, and the United States determined that 76 per cent of consumers watch video on their personal computers, 32 per cent watch video on a portable device such as a cellular phone, and more than half of the online video audience say they are watching less television because of online video.[25] Most video viewing takes place at work.[26] Over 57 per cent of the online video audience will introduce others to the videos they find.[27]

To summarize all these numbers: The use of YouTube among the global population is increasing. YouTube videos are getting longer. Viewers are spending more time each month on YouTube. YouTube videos rapidly migrate across the Internet population because Internet users tend to share what they find with their friends. Online viewing practices are migrating out of the home and office and taking place in more varied locations.

I do not attempt to catalogue the varieties of online video experience, nor do I focus on the most famous videos and YouTube personalities. Instead, *Watching YouTube* provides a series of case studies that epitomize broad social issues arising from mass participation in amateur video production. Herein I also resist the temptation to provide a grand theory that purports to explain all aspects of amateur culture production. After over half a century of intense study we are still sifting through all the contradictory effects of commercial television. Thus, the study of YouTube and online video both benefits from the ground covered by television studies and suffers from the many unresolved issues that arise when we make ourselves and our media the subject of analysis.

In 1982 Ien Ang's study *Watching Dallas* set in motion a new trend in media analysis in which researchers treated the television audience both seriously and sympathetically. As in the case of the blogging phenomenon that preceded the explosion in online video, there is a tendency among commentators to dismiss YouTube and online amateur video as little more than the digital trash of a generation armed with too much technology, too much spare time, and too little talent. Following in the footsteps of Ang's study, *Watching YouTube* begins with the assumption that there is much to be learned from a detailed study of popular culture and trivial tastes. As in the case of Ang's study, one of the main issues addressed herein has to do with the construction of tastes. What makes online video so popular? YouTube is one of the most visited sites on the Internet. Ang's study was an argument against the overstated force of American cultural imperialism as a controlling influence on television viewing habits. Obviously, we cannot claim that the popularity of laughing babies or Lonelygirl15 results from nefarious American control.

The construction of tastes among the Internet audience is due to a very different set of factors and institutional forces than are found among the television audience. Tastes, according to Pierre Bourdieu, are related to the construction and maintenance of class divisions.[28] So what happens to the social order if tastes are no longer so closely controlled by institutionalized influences? Is the formation of taste among the online audience part of a postmodern shift in the nature of popular culture?

When Ang set about exploring the reasons for the popularity of the American soap opera *Dallas* among Dutch women, she had to solicit responses from the audience. Because of the interactive structure of YouTube, the task of studying audience reception and tastes is somewhat simplified. YouTube allows viewers to submit comments to any video and even reply with a video of their own. Thus, we find a rich collection of sociological and anthropological data strewn throughout YouTube, where some videos generate hundreds of thousands of written responses and thousands of video replies. One may argue that using such data simply shifts the analysis from reliance upon simplistic questionnaire data to simplistic user-generated commentary, but in the end there is no such thing as perfect audience data. What is certain is that the character of the Internet has altered the way audience data can be gathered and has changed the type of data that are available. Of great significance is the ability of the online audience to speak for itself, in an unsolicited fashion, in response to what it likes and dislikes.

Watching YouTube is one of media culture's guilty pleasures. It *almost* has the same social esteem as watching pornography (indeed, in many cases it *is* soft porn). It is done with friends, it is done when alone, it is done at the office, at school, at coffee shops. It can even be done in the car. There is little point in trying to list all the types of content found on YouTube, as from theorist Jean Baudrillard to comedian Jon Stewart, from opera to rap, almost the entire lexicon of modern life is already represented on the 'Tube. Along with funny television ads and pirated clips of situation comedies, sports, and movies, a great deal of what can be found on YouTube is home-made videos. It might be someone ranting about celebrities, talking trash, or simply opening a box that contains a brand new techno-toy, but for the most part it is entertainment. In studying YouTube we are squarely in the domain of popular culture. In the early 1900s popular culture was analyzed strictly in terms of an unwitting audience, irresistible ideologies, and morally suspect content. Theorists now see the 'popular' as a field of contested meanings, subversive strategies, co-option, and resistance.

As I probed deeper into the world of YouTube, it became increasingly clear that the phenomenon of online video does not have easily defined boundaries. Unlike the early years of television and its audience, there is no one place to watch YouTube. Watching YouTube bleeds across all social domains, sexes, and classes. The activity of discussing and watching YouTube is without boundaries or formal structure. Millions of YouTube videos can be found on millions of blogs, accompanied by millions of comments. Amateur videos are found on websites, Facebook and MySpace pages, and the evening television newscast. Thus, a study like this one that focuses on the social use of YouTube will necessarily be incomplete in scope, partial as a cultural record, and tentative in its conclusions; for neither the phenomenon in question nor the entirety of its context can be mapped in a single research project.

Amateur video practices were once almost entirely restricted to home-made movies of families at home and on vacation. They were usually made only by the middle and upper classes, as the technology was pro-hibitively expensive. Today almost every cellular phone and camera also functions as a video camera. Thus, the combination of a video camera in every pocket and a billion people connected to the Internet has made this the high age of amateur video production.

The online audience are not passively watching YouTube. Often they can be found actively engaged in what they have watched. This active stance of the audience leaves a rich record of written and video comments. In 2007 over 13 per cent of the online audience posted com-ments about the videos they watch.[29] These comments tell us a great deal about the nature of this new digital medium and about the general spirit of the times. Unlike watching television, many members of the YouTube audience have a different relationship to the screen because they are also *producers* of videos. Thus, the new media audience interacts with amateur videos from its dual stance as producer and as consumer of video. This adds a new dimension to the character of the online audi-ence and is often reflected in the YouTuber's attitude towards the video work of fellow amateurs.

Like those of the television audience, the responses, tastes, and rea-soning of the online audience are intensely irrational and contradictory. Yet in the end, a coherent picture emerges of online audience behav-iour. There is a zeitgeist to the aesthetics of the YouTube experience, one that is structured by the limitations and possibilities of the medium itself. The aesthetics of the YouTube experience, the settings it uses, the styles it embodies, and the emotions it generates reflect the prevailing condi-tions of late-modern capitalism as it slides slowly into the postmodern condition. The surrounding cultural climate, the constraints of corpo-rate mass media, and the ever-present reality of economic and military empires frame the investigation into YouTube. By paying close attention to who is saying what to whom on YouTube, we will arrive at a better understanding of the changes taking place in media culture, in empire, and in our everyday lives.

Mark Poster notes that the concept of the everyday has been central to critical theory since the early 1960s.[30] Poster approaches the category of the everyday through the work of Henri Lefebvre. In 1947 in the *Critique of Everyday Life* Lefebvre wrote that everyday life is a residual category that is defined by 'what is left over after all distinct, superior, structured activities have been singled out … Everyday life is profoundly related to all activities, and encompasses them with all their differences and their conflicts; it is their meeting place, their bond, their common ground.'[31] YouTube is providing us with new insight into everyday activities.

For Lefebvre the everyday is the area of life that exists outside of political and economic institutions and thus represents a space beyond institutional frames of domination. Poster likewise defines the everyday as a byproduct of the institutionalization of activity. Within a modern society organized by institutions, 'whatever is informal, unorganized, serendipitous, and chaotic may be considered the everyday … [the everyday] escapes institutional appropriation by the massive powers of the state and the economy.'[32] In the context of a highly disciplined social order, the everyday video practices of amateurs provide a space for hope, optimism, freedom, liberation, and resistance.

Of course, the everyday is not a domain of complete freedom, nor is it completely free from the surrounding ideological apparatus of commercial media. As the individual is deeply entrenched in ideological modes of thinking, self-knowledge has become problematic within institutionalized mass society. This is not a new issue. The problem of self-knowledge lies at the heart of religions and philosophies both ancient and modern. Marxist, conservative, and liberal theorists regard the individual in the same manner as philosophers, psychologists, and religious scholars – often we do not know our own minds.

Along with Freud, who famously declared that we are not the masters of our own house, Lefebvre saw the everyday as the realm of alienation and inauthenticity. Since everyday life is corrupted by the mass media, self-knowledge is at best imperfect, 'men have no knowledge of their own lives: they see them and act them out via ideological themes and ethical values.'[33] Lefebvre feared that the liberatory possibilities of the everyday were about to be overwhelmed by a totally bureaucratized and administered society. We also find in the YouTube community a fear of impending totalizing control.

Like many contemporary critical theorists, Poster also sees the everyday as a realm of distortion, domination, false consciousness, and all things related to ideology. The notion that there is no refuge from ideology is found among both modernists such as Anthony Giddens and postmodernists such as Poster, who quite rightly describes the everyday as suffering from a 'distracted participation in ideology.'[34] Thus, Lefebvre's definition of the everyday as that which lies outside the control of the institutional apparatus of the state and the economy needs to be carefully qualified. As we all suffer from the effects of the ideological, so too is the realm of the everyday interpenetrated by ideology. Since ideology is never uncontested, this also implies that the realm of the everyday is a realm of resistance. As a domain that represents the mundane realities of everyday existence, YouTube is a cultural field where we participate in ideologies and also express our resistance to domination.

In the context of our information society, argues Poster, everyday life is transformed into 'a battleground over the nature of human identity.'[35] Networked computer systems such as the Internet have expanded the struggle over all forms of identity. As will be seen in the following chapters, amateur online video is bringing more everyday experiences into the arena of struggle over identity.

Today it is hard to identify spaces in everyday existence that are free from media use. During the writing of *Watching YouTube* I took the train from Ottawa to Montreal. In search of a functioning washroom, I stumbled into a section of the train that was filled with young teenagers on a school trip. As I walked down the aisle, I noted the proportion who were engaged in a media-related activity, a total of 40 per cent – well over twice as much media activity as I found in the adult-populated sections of the train. From adults addicted to their Blackberries and iPhones to young people endlessly gaming and checking their Facebook sites, the everyday is deeply interpenetrated by media consumption. As has been shown in so many ways, our lives draw tight circles around our media systems and mediating technologies, so much so that one might well say that media constitute the practice of everyday life.

The most frequently uploaded videos to YouTube are made at home by amateurs and capture aspects of everyday life.[36] For the most part the videographers explored herein are recognizable as amateurs. With few exceptions, they are not part of the commercial media system (although some aspire to be). Many lack talent and some should consider not making videos at all. There are areas of video practice where professional and amateur overlap. Like so many other such distinctions, the category 'amateur' does not have seamless boundaries. Any useful definition of amateur cultural production must recognize that often we have our feet planted in more than one world.

Patricia G. Lange suggests that YouTube is 'weighted towards the non-ordinary.' I am inclined to think that Lange is only half right on the non-ordinary nature of people involved with YouTube, particularly those who make videos. Lange proposes, 'if you are posting videos on YouTube, you are arguably no longer ordinary.'[37] There was a time when one could say the same thing about those who used email, but that time passed so quickly that many can barely remember it (certainly none of my students). Consider that there are 2.8 billion cell phones in use, many of which have built-in video cameras. There are over 130 million new digital video cameras sold every year. These numbers add to the existing count of 1 billion digital cameras already in use. The result is approximately 2 to 4 billion consumer devices equipped with digital movie cameras.[38] By spring 2009 many advertisements for camcorders were making direct reference to YouTube. Cellular phones and YouTube-friendly pocket camcorders were being sold with features such as direct uploading to YouTube. YouTube is now seen as the logical destination for amateur home videos. Even if Lange is correct, any non-ordinary character of creating moving images and uploading them to the Internet will soon pass into the domain of the ordinary. The activity of posting a video to YouTube is indeed unusual, but it is done by many people who are as ordinary as the rain.

At present the total digital universe is 281 billion gigabytes and is predicted to be 2,810 billion gigabytes by 2011.[39] The digital universe consists of all information in all forms that has been created, captured, and replicated in digital form. Content created by individuals (not by corporations) accounts for approximately 70 per cent of this digital universe. The largest part of this digital universe is visual information: images, camcorder clips, digital television signals, movies on DVDs, and digital streams of visual information from surveillance cameras. A key factor driving the rapid expansion of the digital universe is the increasingly higher resolution of consumer recording devices – we are capturing more of our world at ever higher resolutions.

I was quite surprised to learn that amateurs, not professionals or corporations, accounted for the lion's share of production within the digital universe. In 2007 it was estimated that only 14 per cent of YouTube's content was commercially produced videos.[40] Inside and outside YouTube, amateurs are the number-one producers of visual information and digital culture. Most of this amateur digital culture ends up on the Internet. What remains to be seen is the social and political impact of mass amateur cultural production on a global scale in the digital age.

Watching YouTube's exploration of amateur digital culture begins with a look at the analogue equivalent of YouTube – the old home movies our parents and grandparents made on 8mm and 16mm film cameras. The first chapter looks at how technology and culture combined to create patterns in amateur home movies in the twentieth century. The patterns found in analogue home movies provide us with points of comparison for understanding the significance of amateur online videos. A brief case study of weddings on YouTube highlights changes in home movie-making practices. The second chapter explores how the family and the home are represented on YouTube. A rather disturbing picture emerges of uncontrolled children and teenagers armed with cameras and making victims out of other family members. Chapter 3 asks whether we encounter real people on YouTube or whether we are trapped in some form of regressive hyperreality. Video diaries provide a way to interrogate how YouTube intersects with the surrounding culture of reality and confessional television. Chapter 4 looks at the self-representational practices of young women, black women, indigenous women, Barbie girls, fat girls, thin girls, and violent girls. After a century of suffering under the male gaze, women are turning the camera on themselves and telling their own stories. Chapter 5 looks at YouTube as a platform for social networks and takes the measure of some of YouTube's communities. Here we will sit in the virtual village square and watch the comings and goings of the village cop, digital pirates, invasive celebrities, productive fans, haters, and spammers. Some of the ordinary people we will meet are cat-loving engineers, Gary Brolsma of *Numa Numa* fame, guitarist Jeong-Hyun Lim, anthropologist Michael Wesch, and Maelle Claire (one of the original members of the YouTube Generation who was actually born on the 'Tube). No community exists in perfect peace and so it is with YouTube. Chapter 6 ventures into the field of battle and looks into how amateur videographers use their cameras to debate politics, religion, and war. The final chapter steps back from the screen and asks what all this means for the future of the television audience. No new medium ever leaves old media unchanged.

In a few instances the following chapters will explore some rather horrific behaviour. In cases where I describe videos that are clearly destructive of another individual I have refrained from providing any details that might lead to the discovery of the video itself or the individual involved. I have done this in five case studies so as to prevent further harm from being done. In other instances where an injurious or unethical video has already been seen by many people and discussed by the press (such as the case of the *Star Wars Kid*) I have not withheld details, as they are already widely known. On the few occasions where I am quoting the hateful or obscene words of some YouTube members I have also withheld their identity. Herein I would rather avoid unnecessarily drawing attention to individuals who seek attention by heinous means.

A Technically Brief Introduction to YouTube

My purpose here is not to describe how to use YouTube or document all its features and uses, as they are legion. Some odd terms associated with YouTube and used in the following chapters merit a brief explanation for those new to this digital frontier. Any Internet user may view a YouTube video free of charge but only YouTube members may upload or post (as it is often called) a video. Doing so is quite easy and I now require all my students to become members of YouTube and create and post at least one video.

You become a member of YouTube by registering and creating what is called a channel. Registration is free, as is the general use of YouTube. Individuals and corporations have the option of paying for a partner channel that provides special privileges. A channel is a 'page' in YouTube that is similar to a Facebook page (with many key differences). A channel can be customized with images and information and acts as a repository for all videos uploaded by a member. Two of my channels are EmpireofMind and AutoVideography. Many users have no name other than their channel name, so a channel name frequently is used as the individual's name. Channels provide details such as name, age, location – all of which can be true or false, which presents obvious research difficulties. It is often possible to find more details about videographers via Google's search engine or Wikipedia if they have a significant audience.

Because the YouTube corporation can delete a member's channel and all its videos without notice, it is important to keep a backup of videos. YouTube is not a reliable extension of a personal hard drive. It is private property, a fact that many members seem to overlook.

Videos on YouTube can be commented upon with written or video replies by others, unless the individual who 'owns' a given video decided to turn off these features. Videos that are copyright violations or deemed to violate some aspect of YouTube's policies can be flagged. When a video is flagged, YouTube personnel investigate the issue, but this process lacks transparency and is unevenly applied. Many artists, scholars, and lawyers argue that YouTube's policies are biased, haphazardly and inconsistently applied, and harmful to global culture.

Herein, videos are sometimes referred to as clips. The names of

Dr Strangelove's YouTube channel, EmpireofMind.

YouTube videos appear in italics, for example, *Dr. Strangelove Makes a Cup of Coffee* (271 views). Often I have recorded the number of views for present consideration and future evaluation of growth rates. These view counts are imprecise measures. They can be artificially inflated and, for a variety of reasons, they can also be understated. This presents obvious problems for advertisers.[41] Often there are many copies of the same video under different names. Sometimes a highly viewed video is deleted and then reposted, which causes the view count to start over at zero. Since view counts include repeat viewings, they only approximate the actual audience size. In cases of highly popular videos the total size of the online audience can be larger than YouTube's view count by a factor of ten to twenty.

All the videos discussed in any detail are listed separately in the bibliography. As either YouTube or the video owner may delete a video at any time, these videos may no longer be on YouTube. Videos never really disappear from the Internet if they have an audience intent on seeing them. Deleted videos can be found under another name or on other video websites. For those interested in delving deeper into the subject of amateur online video I maintain a comprehensive YouTube bibliography at my blog, *Watching YouTube* (www.strangelove.com/blog). Fellow Facebook addicts can follow my postings on media culture at www.facebook.com/michael.strangelove. My Twitter microblog, www.twitter.com/Doc_Strangelove, tracks industry reports, statistics, and significant news items about online video.

1 Home Movies in a Global Village

The exploration of amateur online video has no necessary point of departure or destination, so like a hobbit's journey into strange and unfamiliar lands, one might as well begin at home. In searching for a place to begin the analysis of YouTube and amateur online videos I find an obvious analogy to the home movies of the boomer generation. YouTube is home to millions of amateur videos. Yet it is not a living room wall or a projection screen. Digital cameras are not 8mm or 16mm film cameras, and the audience is not merely a close circle of friends and family members.

Gone are the days of bulky equipment and expensive cameras, films, and developing fees. Now digital camcorders are small, relatively inexpensive, and universal. Most digital cameras and cellular phones also function as movie cameras. New, powerful 'smart phones' feature one-button wireless video uploading directly to YouTube. Many desktop computers and laptops come with webcams built in or as a free accessory. Low-cost digital movie cameras that even can be used underwater are made for children. Today there are billions of consumer devices that can record moving images.

In one form or another the digital movie camera is a ubiquitous part of the consumer lifestyle. Along with the astonishing advances in low-priced digital cameras, it is now possible to carry a miniaturized projector that is about the size of a cellular phone and can fit in a pocket.[1] Some cameras even have built-in projectors. Production and distribution technology of home movies have undergone dramatic changes. The audience has changed, the viewing context has changed, and so too have our sensibilities about the appropriate subject matter and style of home moviemaking. Home movies are not what they used to be.

The cultural field of home movies overlaps the broader category of amateur films (also known as orphaned films) and both include a wide range of styles and formats. Film historian Patricia R. Zimmerman describes amateur cinema as encompassing 'a variety of ethnographic, industrial, labor, scientific, educational, narrative, travel, missionary, explorer, cine club, art, and documentary forms produced for specialized exhibition in clubs, churches, and schools and on lecture tours.'[2] Indeed, it is difficult to identify any aspect of human activity or style of filmmaking that is not represented within amateur cinema.

One of the main points of distinction between amateur and professional filmmaking is the role of the marketplace. The European organization for amateur film, Association Européenne Inédits, defines amateur films and videos as 'originally not meant for viewing in professional audio visual circuits.'[3] Iván Trujillo, former president of the International Film Archives, defines amateur cinema as 'films without an interest in profit, produced by technicians and actors who are not financially compensated.'[4]

In the past the majority of amateur filmmakers had no aspirations to commercial distribution and fame. Prior to the advent of the Internet the gulf between the amateur moviemaker and the professional world of Hollywood and television was simply too great. Now, given so many recent examples of individuals crossing over from the world of YouTube to the television screen, amateur video can no longer be defined by its lack of interest in accessing commercial distribution channels. Simply put, it is quite clear that many amateur YouTubers are seeking fame and fortune. Like many claims about YouTube, though, this requires careful qualification. Amateur videos on YouTube are also intended for niche communities, family members, or simply friends. Nonetheless, one of the first things to note about YouTube as a mass cultural phenomenon is that it has altered the amateur's relationship to the television and film industries and expanded the commercial motives for amateur filmmaking.

Amateur online video, like many of the cultural practices of the masses, faces an uphill battle for legitimacy and significance among the intellectual and cultural elite. As Zimmerman notes, home movies are 'often defined by negation: noncommercial, nonprofessional, unnecessary.' Home movies are seen as 'an irrelevant pastime or nostalgic mementos of the past, or dismissed as insignificant byproducts of consumer technology.'[5] Amateur online video has also been called irrelevant and insignificant. Media columnist Andrew Keen sees nothing more in YouTube than 'inanity and absurdity ... Nothing seems too prosaic or narcissistic for these videographer monkeys. The site is an infinite gallery of amateur movies showing poor fools' and 'user-generated nonsense.'[6] Ironically, while he claims that the Internet is delivering 'superficial observations of the world around us rather than deep analysis, shrill opinion rather than considered judgment,' Keen himself more or less does exactly that in his rant, *The Cult of the Amateur: How Today's Internet is Killing Our Culture.*[7] Keen's dismissive overstatements reflect the way deinstitutionalized, non-market cultural production has been marginalized.

Zimmerman argues that amateur film is history from below, unexplored evidence, potentially subversive in its meanings and implications, 'a necessary and vital part of visual culture.'[8] She captures the complexity embodied in amateur films when she suggests that they should be analyzed not 'solely as artistic inventions ... but as a series of active relationships between the maker and subject, between the film and history, between representation and history, between the international and the local, between reality and fantasy, between the real and the imaginary, between the nation and the empire, and between gender and race.'[9] The same needs to be said about amateur online video.

Both historians and media theorists have argued that amateur films have redefined national identity, challenged normative assumptions about gender roles, subverted the agenda of corporate news, mediated historical traumas (think of the Zapruder film and videos of the twin towers on 9/11), supported existing social orders, and propagated counternarratives. Yet only recently have we come to recognize that amateur film enables us to know more about the past. It was not until 1995 that the first book-length treatment of the history of amateur film was published in the United States.[10] The time will soon come when amateur videos on YouTube are treated as significant sources of historical insight.

In 1982 anthropologist Richard Chalfen observed that the scholarly analysis of home movies as a form of social communication was 'virtually nonexistent.'[11] Sociologists have also been accused of largely ignoring visual representation.[12] In 1991 Sean Cubitt asked, 'why has "video" now taken on an identity as, by and large, the least respectable of all media?' (Could it be because of its early association with the pornography industry and, more recently, its association with amateurs and the Internet?)[13] As recently as 1998 Zimmerman found that many scholars 'had never considered home movies to be a topic for serious, much less academic, consideration.'[14] In 2008 both an archivist and an anthropologist mused that 'the study of amateur images is still underdeveloped. Serious consideration of amateur images is new territory for most researchers.'[15]

Henceforth, the subject of amateur representational practices will be much more difficult to overlook or dismiss. The mass participation in online videomaking is already having an effect on the way events are recorded and remembered and deserves an important position in all areas of the human sciences.

How we think and how we perceive the world is affected by the technologies we use. Cubitt makes a compelling argument that, as Marshall McLuhan famously observed, there is a relationship between media technologies and psychic structures. This is not simply a private matter between me and my technology. As the viewing experience is a social experience, there is also a relationship between my mind and yours. Thus, the analysis of video requires attention to the institutional, psychic, and intra-psychic dimensions. The challenge for theorists is to keep all these balls in the air as they juggle market forces, the internal realm of Freud, and the external realm of family, friends, and strangers in the analysis of laughing babies and lonely girls on YouTube.

Whereas sociological and literary approaches to film, television, and video have privileged 'the producer and the text,' Cubitt observes that more recent trends among researchers and theoreticians have focused on the audience.[16] This mirrors broader trends in media and cultural studies that emerged in the 1980s and saw an active audience moving to the centre of analysis. Herein I will be following in the tradition that Cubitt's own writing represents and focusing on what people do when they use video. In the words of anthropologist Clifford Geertz, our first task is to determine what amateur videographers 'take to be the point of what they are doing' and interpret their actions, words, and meanings while trying to 'steer between overinterpretation and underinterpretation.'[17]

Home Movies and Patterns of Culture

Along with their obvious significance to historians and media theorists, amateur online videos have caught the attention of anthropologists. Anthropologists have long used film as a tool to capture and interrogate cultures.[18] Chalfen explored the significance of home movies with an anthropologist's concern for patterns of shared beliefs and behaviour: 'we are not examining instances of idiosyncratic picture-making behaviour.' Anthropologists see home movies and YouTube videos as expressions of social norms. Home movies are culturally structured documents that are the result of 'shared knowledge, tacit consensus, informally learned behaviour, models for appropriate behaviour, and structured patterns of interpretation.'[19] Much of the work in visual anthropology on home movies has been inspired by Sol Worth's seemingly simple insight that 'one looks for patterns dealing with, for example, what can be photographed and what cannot, what content can be displayed, what was actually displayed, and how that display was actually organized and structured.'[20] Within YouTube's universe of amateur videos can be found all sorts of patterns. These patterns reflect choices and these choices can tell us much about ourselves.

The patterning process of culture works in two directions. Culture serves to organize how home movies are made and also produces patterns of behaviour captured within home movies. Chalfen proposed that one of the primary tasks in analyzing home movies is to discover a pattern of content and its underlying orderly view of the world. This view, of course, raises questions about the postmodern condition in which the YouTube generation lives, a condition defined by fragmentation, the decay of shared social orders, and the indeterminacy of meaning, which suggests that we may find both cultural patterns and cultural fragmentation within online videos.

Chalfen argued in 1975 that there was a 'limited number of people participating in a confined array of places and events' within home movies.[21] His observation was quite valid at the time. Home movies had a narrow cast of actors, usually immediate family members and close friends, and portrayed a narrow range of settings and events – holidays, birthdays, weddings, homes, cottages, and domestic settings.

When I told my parents that I was writing a book about home movies, my mother went to a closet and dug out two reels of processed Kodachrome colour 8mm movie film – a total of fifty feet (15m) of family memories I had never seen before. I converted the films to video and posted them on YouTube so the whole family could watch them. The old films captured moving images, from before my birth, of my parents, relatives, and a grandmother I never knew. The scenes were typical 'home-movie' material – vintage cars with tail fins, the family returning from church, visiting relatives, parents clowning around for the camera, children and infants playing in the front yard, horn-rimmed glasses, beehive hairstyles, and bathing suits that are, oddly enough, back in style. These soundless movie images presented a happy, playful family. The scenes could have been taken from an old Disney movie. In the future will we discover long-lost home videos scattered across YouTube and the Internet? Probably.

The Slade family home movie, 1959.

Chalfen's description of home movies as a representational form defined by a circle of intimacy, limited settings, and conventional aesthetics
is echoed by visual anthropologist Jeffrey Ruoff: 'Home movies offer conventionalized representations of the world through the cinema. A clearly
defined etiquette exists for the types of images made, the circumstances
under which they are made, and the persons and events represented. In
addition, the contexts of exhibition are highly restricted.'[22] The advent
of online amateur video has changed all of these elements of who, what,
how, why, and where. YouTube represents a disruption in the cultural
patterns of analogue (film-based) home movies. It is introducing us to
new patterns and styles of representation.

The low cost of video cameras, mass participation in the Internet within industrial states, and new social norms of twenty-first-century media
culture have effectively blown the lid off the cast, plot, settings, and styles

of online videos. Potentially, everybody is the subject and anywhere is the location. Nonetheless, it remains to be determined if shared norms and conventionalized representations no longer play any role in amateur videomaking.

Unlike the material produced by YouTube's amateur videographers, home movies were conventional. The institutionalized 'how-to' discourse of amateur filmmaking in the twentieth century had a clear bias against idiosyncratic behaviour. Amateur filmmakers were advised by professionals to 'take a less egocentric approach' to their subject and edit out aspects of their personality which were 'difficult or confrontational.'[23] There were exceptions, such as art school projects, avant-garde experimentation, and marginal groups with their own visual culture, but these remained in the minority.[24] Assumptions about what an amateur video or film should look like were based on the paradigm of television and commercial cinema.

Instructional manuals for videomaking were 'extraordinarily prescriptive and dogmatic.'[25] One of the most persistent and widespread assumptions about the aesthetics of home movies was that they should look natural and authentic. Multiple sources of how-to instructions, such as workshops, manuals, and magazines, propagated assumptions about the proper subject and aesthetics of home movies, which in turn served to push the idiosyncratic from the home theatre. This conformity, in turn, generated homogeneous patterns within the home movies of the twentieth century as anomalous behaviour and acts that lay outside dominant norms were edited from view.

Although amateur moviemakers once avoided recording the idiosyncratic, this is no longer the case. YouTube provides an opportunity for individuals to display highly creative and sometimes highly destructive idiosyncratic behaviour. Once captured by the amateur videographer, these expressive acts are widely copied and become yet one more cultural pattern. This is most clearly seen in the spread of new dances via amateur YouTube videographers. Styles of self-presentation and identity construction are rapidly created and reproduced by online videographers. YouTube has fostered a competitive environment wherein videographers compete for attention and popularity by innovating new styles of idiosyncratic behaviour. In so doing, amateur videos also push the boundaries of norms and challenge conceptions of appropriate behaviour.

Consider the vomit genre of amateur online videos. On YouTube we witness barfing pets, babies, teenagers, and adults.[26] The location of vomit scenes is even more varied than the subjects: homes, airplanes, roller coasters, cars, boats, toilets, in the street, off rooftops, and in shopping malls. A subgenre of such videos depicts individuals drinking Ipecac syrup (an emetic) and then vomiting. In common with similar productions, these videos appear to encourage copycat behaviour among young males. Amateur videos in this genre are influenced by the rise of scenes of vomiting in movies such as *Stand By Me*, animation such as *Family Guy*, advertisements such as E*Trade's 2008 Superbowl commercial (which depicts a baby vomiting), and various television shows. YouTube also captures all these moments through pirated clips. Curiously, vomiting has been found to be the most upsetting and irritating sound, so it is quite peculiar to find an increase in the depiction of vomiting in movies, on television, and within amateur videos.[27]

Within the Internet's rich field of amateur videos we can find other departures from late-twentieth-century patterns in home movies. Chalfen noted that home movies of the 1970s generally did not focus on unfamiliar people, death, arguments, use of the bathroom, vomiting, and sexual intercourse. Amateur home moviemakers 'independently reached a consensus on what segments of daily social life warrant filmic treatment.'[28] The stranger, death, violent arguments, people using the toilet or in the bathtub, children vomiting, and amateur pornography are common themes within contemporary online amateur video. As a corporation YouTube exerts censorial control over what type of content may be hosted on its server, but this has only a marginal effect on the range of content found within amateur videos. Within the Internet community there is no such thing as consensus over content, cinematic or otherwise. We may be close to the point where all forms of behaviour are represented online.

Chalfen also suggested, '[the] home-movie pattern has been much more consistent through time than we discover for feature film.'[29] Yet this home-movie pattern has been broken by the YouTube phenomenon. The old patterns of who (family and friends), what (small slices of life that excluded many aspects of day-to-day existence), and for whom (an audience composed of family and friends) no longer dominate home movie production choices. It may well be the case that consistency is now much more characteristic of Hollywood than of YouTube's global army of amateur videographers. Hollywood produces formulaic movies as a result of its production methodologies and economics, whereas amateur videographers are freed from such institutionalized constraints. The patterns of Hollywood films, their 'iconic visual spectacle and rhetorically simplistic individual discourse,' help to legitimate dominant political and economic institutions and their value systems.[30] Thus, it may be the case that the mass involvement of amateurs in the production of moving images may help to undermine Hollywood's legitimation of key institutions.

The state and corporations have always policed the reproduction of images. This has led to the homogeneous character of American movies and television shows (a situation that is not without exceptions). One only need explore foreign films at a local video rental outlet to discover styles of storytelling that depart from Hollywood's menu of choices. In contrast to the mass democratization of moviemaking that YouTube represents, Cubitt could rightly note in 1991 that the entire medium of video 'can now be considered as largely devoted to the distribution of Hollywood cinema films.'[31] In the 1980s five major distributors of videos successfully limited the market reach of independent distributors. This strategy of market consolidation set severe limits on the diversity and choices available to the mass audience. The situation is quite different for the audience of amateur videographers. Market forces such as distribution oligopolies and trade agreements have no impact upon the production and distribution of amateur online movies.

Cubitt makes an intriguing suggestion that new reproduction technologies did not lead to a significant break from the patterns of representation established by major commercial studios. Perhaps having internalized the aesthetics of homogeneous programming, 'those who try to appropriate each newly available technology for new purposes seem constrained to reproduce the patterns of textual reproduction which the medium seems to demand.' Cubitt describes some of the constraining forces as the 'tyranny of "good taste," of professional standards, of chasing festivals, prizes and reviews' and 'commonly held beliefs as to what things should look like.'[32] Thus, even after the independent producer escapes the direct control of the state and the market, there remains a deeply entrenched compulsion to conform to dominant modes of representation. Here we see the depth of the connection between industry and psyche.

The relationship between self and society raises the question of amateur online video as a divergent or a conformist practice. Obviously, we can expect to find traces of industry practices and normative aesthetics in amateur video. But it may be the case that mass participation in online videomaking will mark the beginning of a break from dominant patterns of aesthetics that guide storytelling. The new era of online amateur video has already broken with the patterns of home movies found in the twentieth century. A dramatic departure from the patterns of representation established by major commercial studios is not beyond the realm of possibility and appears to be well under way.

Also of note is Chalfen's assertion that home moviemakers revealed a 'persistent belief in the objective power of the camera' – they saw their films as 'untampered, unmanipulated visual recordings of real life.'[33] In the 'low technology' world of film-based (analogue) movie cameras, complex editing and special effects were rare. In contrast, today's domestic video cameras and editing software allow for substantial editing. Some of my students even go so far as to use blue screens for inserting digital backgrounds and special effects into their class projects.

The advanced technology of video has created an audience that is well aware of the amount of fakery that is possible and has led to numerous amateur video genres that attempt to fool the audience. Thus, we must question whether or not the YouTube audience holds the same degree of belief in the objectivity of the camera. We also see how the audience of home movies has changed when Chalfen describes the intended audience as 'people either related to or known to the film-maker.' On YouTube your audience is any of the 1 billion Internet users. All in all, Chalfen's description of home movies as the 'simplest situation' for ethnographic exploration, one in which the filmmaker, the audience, and the subject matter all are similar, demonstrates how very different the situation is among the YouTube generation.[34]

The Demise of Context and the End of Simple

The once simple situation of home movies is further complicated by a characteristic of the Internet that postmodernist media theorist Mark Poster explores extensively in *Information Please: Culture and Politics in the Age of Digital Machines*. Poster describes how the Internet's introduction of a new level of global communication promotes transcultural confusion where misunderstandings abound. The Internet promotes conditions of increased misunderstandings and misinterpretations because cultural objects, such as home movies, are 'disembodied from their point of origin or production.' Home movies were once locally produced and viewed. The conditions that determined their reception and meanings were embedded in this highly localized context, but the Internet is tearing cultural objects out of such local contexts. Perhaps Poster indulges in postmodern excess when he proclaims that 'there is no longer any local soil on the earth,' but he does have a point.[35] What is the 'local' in the digitized global village?

A loss of context can be seen at work in the one-minute video *Brother and Sister* (1,070,958 views).[36] The video depicts a young adult, Wesley, sleeping in a bedroom. A young woman enters the room, jumps on the bed, and teases Wesley. She playfully pulls the blanket off him and tugs at his short pants. Viewers' written comments show that the audience is uncertain if the young woman is his sister or his girlfriend. The videographer, Wesley's sister, replies in the comments: 'This is not me dumbasses. It is my friend.' Many of the comments are quite obscene and focus on the issue of incest.

Many YouTubers make humorous videos that are intentionally out of context. These videos play with meaning by taking scenes from a television show, movie, or newscast and intentionally editing them so they are out of context. Occasionally, though, we see videographers pleading with the YouTube audience not to take their videos out of context.[37] Also, it is not unusual to see YouTubers engaged in an exchange of videos where they accuse each other of taking things out of context. Much of YouTube operates as a commentary on miscommunication.

The circle of intimacy surrounding home movies of the analogue era created a highly restricted context for their exhibition and interpretation. After reviewing some of the aesthetic traits that are peculiar to analogue home movies, such as the lack of editing, character development, and narrative, Ruoff suggests that 'these traits function perfectly well in their proper context; home movies are typically produced by, for, and about family members and friends. Home movies and family albums call upon contextual information to produce meaning. To the intended audience of family and friends, the significance of these documents is readily apparent, whereas they may appear repetitive or banal to outsiders.'[38] The decontextualized environment of the Internet suggests that we may need to rethink the relationship between context and meaning. For example, claims made in the 1980s that images 'must be viewed in the context of their original rhetorical function, as part of the larger discourse in which they originated, in order to understand their intended meaning' may be correct but are unhelpful. Contexts, origins, and intended meanings often are hard to come by in cyberspace.[39]

The loss of context within the globalized environment of the Internet is not a simple matter of context ensuring the truth of the home movie and the Internet eroding that truth. Visual anthropology stands in accord with the general consensus on representation found across all disciplines: 'The anthropology of visual communication undermines the assumption that visual documents provide a reliable, not to mention objective, portrayal of social life.'[40] A picture may be worth a thousand words *and* it may also produce a thousand different truths.

Along with a loss of context and the erosion of meaning there may be a loss of the value or sacredness attributed to digital home movies. They have become common elements of a family's collection of objects. In the analogue era of film, home movies were regarded as treasured family possessions.[41] With the switch from analogue to digital cameras, many home movies never make it off the camera's memory card. Memory cards fill up; sequences and scenes are erased or overwritten by the next recorded event. Although the evidence is anecdotal, I also suspect that, even after they are transferred to computer hard drives, USB memory sticks, compact discs, and video discs (DVDs), digital home movies are far more frequently lost than their film-based predecessors were. It may be that the expense and rarity of film-based home movies ensured that they were more carefully preserved and passed on from generation to generation.

There may be a hidden paradox in the proliferation of digital cameras and the vast increase in home moviemaking. While we are creating more memories, we are also losing more. Perhaps Walter Benjamin was right about the loss of aura that comes with mechanical reproduction.[42] Could it be that digital technology's increased rate of memory reproduction brings in its wake a loss of the sacred character of recorded artefacts?

What was once a simple situation of common patterns, local realities, closed audiences, and naïve epistemologies is now a complex field of globalized cultural production. Whereas much of the home moviemaking of the analogue era was guided by a desire among filmmakers to show themselves to themselves, we now stand witness to a growing compulsion among online videographers to show themselves to the world. In 1973 Worth looked into the future and suggested, 'It is not unreasonable to expect that the New Guinea native, the American Indian, the Eskimo, the peoples of developing and developed states in Africa and Asia, as well as segments of our own society, will soon be able to make moving pictures of the world as they see it and to structure these images in their own way to show us the stories they want to show each other.'[43] For individuals within all these social groups, the use of video to tell their own stories on the Internet is an increasingly common affair.

The similarities and differences between the era of film-based home movies and the current era of digital movies can be demonstrated through a brief exploration of amateur online wedding videos. The Internet has turned the amateur wedding video into the meeting

We know a bride when we see one.

place between the idiosyncratic self and mass patterns of style and storytelling.

Wedding Videos in the Global Screening Room

Weddings are a highly ritualized event. Everyone knows their place and their roles. Regardless of the young bride's determination to put her unique stamp on 'her special day,' weddings in any one cultural region tend to look very similar. The wedding genre provides an opportunity to explore how amateur online videos display highly ritualized, rule-governed behaviour while also displaying highly idiosyncratic behaviour that defies easy categorization within homogeneous cultural patterns. Perhaps one of the main functions of YouTube is that it challenges our sense of the 'normal' by confronting us with the extreme diversity of human behaviour. YouTube often leaves me thinking, 'I've seen six rabbits in a tree, but I ain't never seen that before.'[44]

Cubitt prefaces his analysis of wedding videos by noting that video media 'can only be understood as the product, the site and the source of multiple contradictions, lived out in multiple practices, caught up in multiple struggles.'[45] Weddings are bounded by rules and are also the site of conflict over rules (as any bride or groom is likely to discover). Rules and expectations also shape the professional videographer's work and ensure that every shot reflects general industry standards of precision and skill. The rigorous demands of the profession and the normative expectations of clients give rise to a tension between the standard formula of wedding videos and the need to make every video appear personal and unique. Thus, wedding videos reflect the general predicament of the individual in mass consumer society. Our attempts at expressing unique individuality through standardized products and services leave us caught between the particular and the universal.

Starting with the very reasonable insight that wedding videos link the particular and the universal – it is my unique wedding, but its formulaic representation refers to all other weddings – Cubitt makes a rather curious series of leaps to arrive at the conclusion that the viewer's reaction to the video is predetermined. He presents a peculiar picture of the power of the video over the viewer, a power that determines the viewer's experience of the video. The viewer is said to feel guilty and excluded by watching such an intensely personal and private affair. The particularity of any wedding video is 'so specific it overrides our culturally sanctioned pleasures in watching screened images.' This is not an active audience, but one whose position 'could scarcely be more definite and hemmed in, more surrounded by ritual responses and respect for the authenticity of the experience on screen: few other media experiences so completely rule out the possibility of criticism.' The audience's reception of the video is definite, hemmed in, ritualized, and void of criticism because the wedding video itself is a highly rule-based, formulaic mode of representation. The new position of the video in representing weddings is said to have acquired 'that religious aura which stops us from criticising wedding ceremonies.'[46]

Cubitt refers to psychoanalytic notions of the ego ideal, modes of consciousness acquired in childhood, and an internally fragmented psyche to explain our response to wedding videos. All too typical of video and film criticism, here the theorist engages in a psychoanalytic analysis of an abstract audience without any actual concrete data to back up the claims about how 'we' respond to the media text in question. Such psychoanalytically informed analysis of an abstract audience is found throughout film and video studies.[47] This pattern of interpretation suggests a need for a meta-theory of film and video criticism that explains the omniscient position of interpreters who feel no need to validate their often idiosyncratic insights into the mental states of the audience beyond referring to various psychoanalytical theories. One all too rarely encounters what Geertz calls the 'creed of the "perhaps"' in all-knowing pronouncements about the structure of the audience's consciousness and its interpretation (or, more often, misinterpretation) of the meaning of this or that film, television, or video text.[48]

Nonetheless, Cubitt is insightful when he draws our attention to the intensely ritualized nature of weddings. His observation is well attested to by anthropologists, everyday experience, and the formulaic nature of wedding videos. This ritualized nature provides us with a point of departure for investigating the other side of any cultural pattern – the anomalous – that which does not fit. Amateur wedding videos capture many familiar patterns while also displaying the idiosyncratic and, contra Cubitt, they invite extreme criticism from the online audience.

Amateur videographers, notes Trujillo, have played 'an important role in generating local cultural awareness ... they provide important cultural information about certain places and periods that rivals that found in more professional films.'[49] YouTube may represent one of the greatest advances in bringing more local knowledge to an audience steeped in mythologized imaginaries of Hollywood. A whole new genre of television shows such as *Rich Bride, Poor Bride, Wedding SOS*, and *My Big Redneck Wedding* have brought the semi-private domain of weddings and wedding planning into our living rooms. These shows focus on the particular personalities and desires of the bride and groom while underscoring the familiar patterns of people, places, and things that constitute the marriage rite. All the 'actors' are amateurs, filmed by a professional crew that produces a polished product according to the technical standards and aesthetic expectations of the commercial television industry. When we explore amateur-produced online videos of weddings, we encounter many of the same rituals, but very different settings and actions are revealed.

The amateur video *White Trash Weddin – what the Hell's a YouTube?* captures many of the familiar patterns of an American wedding. The 1-minute, 16-second clip is unedited and seen in very low resolution. The occasion is a mock wedding (a fake ceremony), but it nonetheless reveals much about the local culture of the participants. The mock element is established by the officiant's opening words: 'We are here today to join Jed and Nancy in unholy matrimony.' The bride appears to be pregnant and is wearing a white baseball cap with a veil attached, a white, sleeveless T-shirt, and a cream skirt. The officiant directs the action of the ceremony. All the bridesmaids wear dresses and carry flowers. The bride and groom, along with their attendants, face the officiant. The outdoor ceremony takes place in the summer of 2007 in a suburb of Buffalo, New York, in the backyard of a small single-storey house.

The mock wedding stands in contrast to the normative mode of wedding ceremonies, which is characterized by formality of location, speech, and attire. The groomsmen are wearing baseball hats and sleeveless T-shirts and carry hunting rifles (perhaps shotguns). The groom is wearing a sleeveless T-shirt and green hunting pants. All the spectators are dressed in casual summer wear: T-shirts, caps, and short pants; and some are holding beer cans. The well-known instrumental composition *Dueling Banjos* plays in the background as the officiant speaks. Thus, the clothing, the officiant's speech, the props (guns), and the music all reinforce the theme of a 'shotgun wedding.'

This mock wedding was undertaken with a sense of humour.[50] We hear the officiant encouraging participants to put their teeth in for all pictures, sleep over in the 'guest houses' (by which is meant old abandoned cars parked in the backyard), and slaughter any gifts of livestock before presenting them to the bride. At one point someone is overheard saying, 'this is going to be on YouTube. I know it'; to which the cameraman replies with mock ignorance, 'what the hell's a YouTube?' The cameraman's comment elicits laughter from all.

Mock weddings are part of a larger North American dramatic tradition.[51] More than simply a parody of the liturgical wedding, here we see a parody of what is known as white-trash or redneck culture. Typical of many online amateur videos, *White Trash Weddin – what the Hell's a YouTube?* demonstrates how meanings become indeterminate when context is eroded. As they move from the privacy of our living rooms to the public screening room of the global village – YouTube – home movies often become the target of contempt, hatred, and ridicule. *White Trash Weddin* was viewed over 36,000 times and garnered forty comments, many of which are quite nasty. Comments within YouTube often lend authority to Pierre Bourdieu's suggestion that 'aesthetic intolerance can be terribly violent.[52] It is noteworthy that one of the viewers of *White Trash Weddin* compared the video to the Country Music Television program *My Big Redneck Wedding.* This comment reflects the fact that our experience and knowledge of the codes and conventions of commercial film and television inform and influence our experience of viewing amateur online video.

Cubitt's observations about the relationship between the particular and the universal in wedding videos hold true in this instance. The particular character of this dramatic performance and this instance of local culture stand in contrast to the universal properties of a North American wedding. The existence of the universal (an officiant, bridesmaids, groomsmen, guests, a white outfit, flowers) makes the idiosyncratic character of the particulars all the more visible. Geertz writes that the process of coming to an understanding of a people's culture 'exposes their normalness without reducing their particularity.'[53] But the universal, the very normalness of this group of people having fun, does not guarantee understanding. The online audience is distinctly hostile towards this mock wedding party.

Cubitt claims that weddings are a genre of home video that completely rules out the possibility of criticism. But here we have examined a *mock* wedding video (although many of the negative comments came from YouTubers who thought it was indeed a real wedding). Does the Internet audience respond differently to home videos of real weddings? A quick survey of real weddings on YouTube leaves little doubt that wedding videos are often subject to intense criticism and distasteful comments.

Examples of other amateur wedding videos that invite extreme audience criticism are *Redneck Wedding of the Year*, *The Waffle House Wedding*, *Redneck Wedding*, and other variations on these titles. Videos that depict unusual weddings tend to receive very negative comments, while more normative weddings tend to receive positive comments. This may be the result of different YouTube audiences being drawn to different types of wedding videos. One thing is certain; as the global village joins us in our homes and watches videos of our private moments and special days, we open ourselves up to hostility and hatred as well as to admiration and respect from complete strangers.

YouTube as a Repository of Shared Experiences

Home movies have a special relationship to place. Nico de Klerk describes this particular quality in the following terms: 'Of all types of film, home movies are the examples par excellence of the situational rootedness of filmmaking, because they comprise the only genre close enough to daily life to register ordinary events and interactions, including its moments of posing and staging.'[54] As was seen in the case of the video *White Trash Weddin,* home movies stand to lose coherence and meaning when they are separated from their original context. De Klerk asks, 'what, then, is the significance of home movies outside their natural home?' – a question that YouTube renders particularly pressing. Does a home movie lose its status as a repository of shared experience, as Eric de Kuyper proposes, and become reduced to a more general source of historical knowledge?[55] I think this is an overstatement, as can been seen in the responses to the YouTube video *Disneyland Home Movies.*

Disneyland Home Movies is a home movie of a family's experience of Disneyland, shot sometime in the late 1960s or the early 1970s. This seven-minute clip was viewed on YouTube over 65,000 times and garnered 154 comments, all of which were positive in nature. The comments make it very clear that this home movie functions as a repository of shared experience. In this regard, msjenie's comment is fairly typical: 'Thanks for sharing this video! It brings back great memories from when I was little and went to Disneyland!!!' The video also functions as a source of historical knowledge for its viewers, as is seen in StevenMSchubert's comment: 'wow i love history and this video is full of history.' The video also provides individuals with a way of confirming the accuracy of the memories of others, as is seen in bethany4588's reflection on how it matches her parents' recollections: 'my parents always always talk about how Disneyland use to be, and thanks to you I can see it, and it's just exactly how they have described!! (we are Disney maniacs as well heh :).'

It is striking that the comments on the *Disneyland* home movie are entirely positive, while many of the comments on the *White Trash Weddin* home movie are negative. One video is an iconic memory of a mass corporate culture that lies at the heart of American nationalism. The other is a representation of a denigrated subculture that lies outside the American mainstream. Surely this contrast says something about how the online audience relates to the universal and the particular. In some instances universal experiences are widely embraced and alien particulars are widely rejected.

Summary

The use of YouTube to distribute home-made movies represents a break with the motives and the aesthetic norms that guided the boomer generation's moviemaking activity. Throughout the last century home moviemaking was largely a domestic, private activity of the economically privileged. This meant that home movies were shaped by the dominant cultural practices, ideals, and sensibilities shared by a narrow slice of the population. As a result, home movies had a common look and feel.[56] Amateur moviemaking was guided by sanctioned aesthetic norms and particular notions of control and skill.[57] These norms were propagated by manuals and popular magazines and, as a result, there was 'a great deal of similarity in the home movies produced by members of the same society.'[58]

As home-movie practices moved out of the living room and became the home *videos* of YouTube, we have witnessed a dramatic change in subject matter, styles of representation, audiences, and reception. The global Internet brings an end to the era of the simplest situations of analogue home movies, destroys context, and complicates the reception and interpretation of home videos. Our YouTube videos often capture the particular and idiosyncratic character of ourselves and our local culture while also reminding us of the universal nature of our everyday lives. The online audience often identifies with the presentation of the universal while recoiling from local cultures that are not part of the mainstream. Thus, YouTube acts as a repository for idiosyncratic behaviour, local culture, shared experiences, and collective memories.

YouTube presents a peculiar set of problems for the construction of identity. It strips our memories of the context that gives them meaning and opens up our lives to misinterpretation. It tempts young people to bring the world into their bedrooms when it might be better to keep the door shut and the camera off. At times, amateur video can ignite an intense aesthetic intolerance among some of the online audience. YouTube invites one to open up, to broadcast oneself to a global audience, but gives no guarantees and little protection for the peculiarities of local culture and the private person.

2 The Home and Family on YouTube

The first studio of this mass age of amateur video is the home, and YouTube's first generation of videographers are providing us with new versions of Ozzie and Harriet, Ralph and Alice, Lucy and Desi. YouTube provides us with a window into the home and the changes that are occurring in domestic life. Surfing YouTube's collection of amateur videos is like wandering in and out of the homes of countless strangers. Front yards, backyards, living rooms, kitchens, basements, garages – every nook and cranny is laid bare. The domestic geography of the home forms part of the unconscious of this new mode of mass moviemaking. Mundane and often cluttered, the physical arrangements of furniture, decorations, piles of laundry, toys, and technology can be seen in the background of many videos. As Lev Grossman told readers of *Time* magazine, 'You can learn more about how Americans live just by looking at the backgrounds of YouTube videos – those rumpled bedrooms and toy-strewn basement rec rooms – than you could from 1,000 hours of network television.'[1] The home as a backdrop is taken for granted by amateur videographers. This backdrop is rich in information about suburbia, consumer culture, and family relationships. Historians, psychologists, anthropologists, sociologists, and media scholars are bound to find a wealth of data in the domestic settings of our YouTube lives. This chapter will explore some of the ways that amateur online video is being used in our homes.

Amateur media practice in the home is often labelled as something inferior, such as mere consumption, narcissistic fetish, or the reproduction of ideal notions of the family.[2] YouTube's amateur videographers have been painted as a source of lawlessness, a crisis of expertise, and a collapse of cultural value.[3] These extreme judgments suggest that some critics reduce the significance of amateur video to mere deviant representational practices.

Amateur videos often capture significant information in the domestic background.

The essence of home video is usually defined according to technological or aesthetic characteristics that are thought to be unique to the medium. These properties often turn out to be shared by other media. Thus, claims about the superiority of one medium over another or the essential differences between technologies of representation are highly suspect. Differences do exist, but we need to pay attention to how groups with vested interests make claims about media practices. James M. Moran suggests that various media practices, such as amateur moviemaking or Hollywood cinema, are used in the social world to 'legitimate invidious social distinctions.'[4] Different types of media practice signify different positions within social classes and different degrees of power and privilege.

The most obvious difference in modern society is defined by the marketplace. Amateur practices exist largely outside the market and are generally lacking in the social power and privilege that come with professional market-based practices. Moran suggests that home video's dominant style of realism and its primary subject of the family are seen as suspect in academic, art, and museum cultures that elevate media activism or the aesthetics of the avant-garde over the mundane practices of home video. Moran points out that, when video technology began to replace Super-8 filmmaking among amateurs in the mid-1980s, academics, journalists, and purists denounced the new technology as artless and corrupt.[5] Oddly enough, in the decade prior to the mass adoption of Internet video there were thirty-six American feature films that presented video in a hostile light.[6]

Amateur Online Video: Defined by Use

Moran proposes that video should be defined by how technology and practices of use combine with human desire and intention. What is specific to video, what distinguishes it from television and film, arises out of how we are using it now: 'Distinctions among media are culturally constructed, and irreducible to the empirical.'[7] What constitutes video is conditioned by the unique circumstances of each moment. In this, the home looms large.

Many popular amateur videos, some of which draw audiences measured in the millions, are set in the home and focus on family members. Numerous professionally created YouTube characters, such as Lonely-Girl15, are also situated in the home. Of amateur videos that appeared in YouTube's category of 'Most Viewed (All Time)' 67 per cent featured the home or the family as the location or focus of the video.[8] There is a powerful relationship between online video, YouTube, its audiences, and the domestic setting. The preponderance of this setting in YouTube merits explanation. For over fifty years family snapshots, home movies, and home videos have been the most common form of amateur media practice.[9] Amateur online video is used in many more varied places than the home, yet the home and family nonetheless dominate its storylines and settings.

This chapter will draw on Moran's study of home video and explore issues that arise from the dense configuration of home and family in the world of YouTube. His study presents a convincing argument that home video practice maintains a close relationship with home movies and television's domestic situation comedies. He provides an important correction to the work of two of the most influential scholars of home movies, Richard Chalfen and Patricia Zimmerman, who focused on the high age of the nuclear family during the decades that immediately followed the end of the Second World War.[10] Chalfen and Zimmerman's models interpreted contemporary home video practice as 'either an insignificant variation of home movies or their illegitimate offspring.'[11] There are reasons for seeing amateur online video as neither insignificant nor illegitimate.

Chalfen's ethnographic research focused on how amateur practices of home moviemaking were standardized by technological and cultural forces. Overemphasizing the standardization of the social world was a common mistake of mid-twentieth-century theory. Zimmerman's historiography focused on how capitalism's ideology and a normative version of the nuclear family shaped amateur practices of home moviemaking. An overemphasis on the power that ideologies had over individuals was characteristic of Marxist-informed analysis in the mid-twentieth century. Both theorists interpreted contemporary forms of family representation through these theoretical frames and saw standardization and ideology as significant factors within home videos.[12]

Introducing the Family to YouTube

There is a genre of YouTube videos that introduces the viewer to an individual's family. These videos often appear with titles such as *Introducing Our Family*, *Meet My Folks*, *Meet My Parents*, and *Meet My Family*. This genre highlights the fact that one of the differences between home movies and digital videos is the intended audience.

In the era of analogue home movies the audience was almost entirely restricted to other family members. In the digital era often the entire YouTube community is the intended audience. We see an example of this in a three-minute video titled *Meet My Parents* ... made by YouTube member Sharron Rosa Giles. Giles takes us to the front door of her parents' house and then inside. Mother and father are invited to 'say hello to YouTube.' Giles then turns the camera on herself and directly addresses the viewer. She explains her relationship with her parents as a rebellious teenager and how time has given her a deeper appreciation for them. She asks the audience, 'How many of you on YouTube still have their mom and dad and do you have a really good relationship with your parents?'

Viewed only 134 times, this touching slice of a stranger's life gives us

Ordinary families and everyday behaviour are subjected to intense documentation due to the rise of digital cameras and YouTube.

a tiny glimpse into an elderly couple's evening as they sit in front of the television with tea and milk on a tray between them and a cat on a lap. Through the video we encounter a stranger who has come to appreciate her parents. Her experience reminded me of my own growing up and developing a more meaningful relationship with my parents. The video creates a bridge between the particular experience of one individual and the universal experience of family relationships that can change over time.

Amateur videos that directly address the Internet audience are common within YouTube. This globalization of the intended audience may be a distinguishing feature of amateur online video. The presence of the Internet has added a new intent to the impulse behind home videography – the intent to speak to a global community. This intention may have existed in the analogue era of home movies, but it was not actualized to the same degree.

Although few amateur videos gain audiences measured in the millions, we nonetheless frequently encounter the desire to address the YouTube audience directly. All manner of performance artists, celebrity seekers, student projects, video works by teachers and researchers, and video diarists speak directly to YouTube. It can be said that a new 'imaginary' exists in the minds of amateur videographers – the imagined global Internet audience. No longer confined to being seen in the family room, home videos can address the online global village.

Conventional perceptions of where the audience is to be found and who its members are have been altered by the rise of YouTube. This confirms Moran's speculation that home video's new technological capabilities may lead to 'a greater range of social intentions' that inspire the use of the camera.[13] YouTube embodies a new type of audience and gives rise to new motives for videographers.

Giles's video is more than a one-way global address. It also resembles one of the oldest forms of Internet communication, the Request for Comments (RFCs). In 1969, when what was to become the Internet was known as ARPANET, Steve Crocker began organizing technical notes from ARPANET engineers into RFCs. These documents captured and systemized the dialogue of network engineers as they developed standards and protocols. RFCs also stand as historical records that documented the debate over various design choices and their impact on the Internet's communicative capabilities.[14] The requests for comments reflect the dialogical character of Internet cultural production. Video also operates as a form of request for comments and indicates a desire for conversation.

Videos act as requests for comments and dialogue in a variety of contexts. Sometimes we see videographers opening up a dialogue with the Internet community so as to explore the mutual experience of YouTube. Adam Quigley's video *How does the world feel to you through YouTube?* asks fellow YouTubers to describe the emotional impact of viewing the world through YouTube. Highlighting the different audience experience that YouTube creates, matry1d responds, 'I just find it awesome to see the comments that people leave on the things they choose to watch ... you get real opinions from real people who choose what they want to watch as opposed to having the TV tell you how the world feels about certain things.' Here we see how a member of the YouTube audience experiences viewing choices, opinions, and the people in the videos as more authentic than those of television.

In another text comment, nocadgem reflects on the conversational nature of YouTube: 'YouTube is a forum where conversation may begin and interpretation follow. It opens a relatively safe space for bridges to be built and connections made across geographies that may other-wise not allow for such interactions and exchanges.' While this viewer is aware of 'the abundance of crap' on YouTube, he nonetheless finds that the YouTube experience 'has a surprisingly positive effect on the way I feel about the world. It may sound cliché, but for me, it really does make the world feel smaller and even more accessible. I think the most exciting aspect of YouTube is dialogue.' Arguments have been made that YouTube is more of an entertainment site than a venue for dialogue, but this conclusion overlooks the high volume of conversation that takes place there.[15]

Dialogue occurs within the comments section of YouTube videos and across a wide spectrum of video genres. Numerous videos complain about haters and bullies on YouTube and try to promote dialogue on the issue. This type of issue-oriented conversation is common within YouTube. Another form of dialogue is 'reply to comments' videos in which videographers directly address their audiences and reply to previ-ous responses to their videos. In these videos amateur videographers are often seen to justify their use of YouTube and defend their ideas and values. YouTubers also make videos that function as responses to other videos. These can engender further responses, and thus we see videos with titles such as MsDiscord's *Response to a response to a video :D*. The conversational character of YouTube is part of the reason why the site is so attractive to its users. Teresa Rizzo suggests that a system that operates on the basis that subscribers upload clips with the hope that they are noticed, commented on and even shared is a system that is based on an aesthetic of attraction.[16] In other words, participation in YouTube can be highly seductive.

Henry Jenkins has observed how the analysis of YouTube videos often treats videos and videographers in isolation.[17] In common with many blogs, numerous amateur videos are isolated and also fail to attract viewers and initiate conversations. Yet, in general, an amateur video on YouTube should be analysed not merely as a text but as a process. Cul-turally significant amateur online videos engender a substantial volume of written comments, video replies, appropriations, and parodies. The meaning and significance of an amateur video are found in how the community responds to it.

One of the results of the conversational character of amateur video practice is the creation of a YouTube community identity among its users. Like rebellious teenagers who construct their identity by differentiating themselves from their parents' identity and values, amateur videographers define the identity of YouTube as 'a creative alternative to television, a place for user-generated content and as a community with norms and rules of behavior that should be acknowledged and followed.'[18] Whether it is posting one's thoughts to a blog, playing an online game, creating a website, or posting a video on YouTube, the act of putting content on the Internet often results in dialogue. One might go so far as to say that, in moving to the distribution platform of the Internet, home videos have been transformed from moments of representation to acts of dialogue. Within YouTube amateur videos often take the form of requests for communication.

The Candid Home Camera

One of the main reasons why YouTube became so popular so fast is because it offers communication and virtual community while also providing less idealized, less polished, versions of ourselves and our world. Attractive aspects of YouTube are dialogue, access to opinions and people, and the sense that we are seeing things that cannot be seen within the regular fare of commercial media. Idealized representations of the home do exist within YouTube, but overall it seems to present a grittier representation of home life.

Videos with titles such as *My Parents Drunk, Drunk Mom, Drunk Dad, My Parents Hate Me, My Parents Fighting,* and *My Stupid Family* show how the ubiquitous presence of video in the home has moved domestic representational practices from eliminating family flaws to including family feuds. The movement from exclusive home movies of the past to inclusive home videos is partially due to the expanding role of teenagers and children in home video practices. Whereas the home movie camera of the past was most often operated by the father, the ubiquitous video camera is found in the hands of all family members, including teenagers and young children. In some cases people have attached a camera to the family dog or cat for a 'pet's eye' view of the home.

Amateur home videographers are recording intimate and highly embarrassing family moments. There are thousands of videos on YouTube that feature drunk parents. It is safe to assume that many of these parents do not know that their children have posted these videos on YouTube. One video clip posted in spring 2008 depicts a mother apparently in a drunken stupor. This two-minute video depicts a grey-haired, middle-aged woman unconscious on a couch in a living room. A young girl (approximately eleven to thirteen years old) shakes the mother's head back and forth, pulls open her eyelids, slaps her face and tries to wake her. Another child holds the camera. The action is neither gentle nor violent. The young girl and the unseen individual holding the camera are heard laughing and giggling. After a minute of this the mother wakes up. The next scene depicts the mother smoking on the couch and conversing with the children in a slurred voice about how she was talking with God. In the description attached to the video the videographer explains how the camera was destroyed a few days later by the mother 'in a drunken rage.' This video was viewed over 10,000 times and elicited thirty-nine comments. In the comments the children and the mother are targets of sympathy and outrage, respectively. This is reality in the raw seen through the eyes of children.

In another YouTube video (19,437 views) the videographer pushes a bathroom door open to reveal an intoxicated mother sitting on a toilet. Individuals off camera are laughing. A young male voice asks, 'are you wasted?' and the mother answers, 'yes.' The mother grabs a bath towel to conceal her face. The videographer, a teenage boy, turns the camera to the bathroom mirror, briefly revealing himself to the viewer, and then says, 'Your mom is trashed: how does this make you feel?' We catch a glimpse of a teenage girl in the next frame. The mother yells, 'Go away; I am going to kick your ass,' as she struggles to pull up her panties and keep the towel over her head. A young girl (apparently the daughter) says in a slurred voice, 'we're going to put this on the Internet, put this on YouTube,' and laughs. Finally, the mother calls out (to her husband perhaps), 'Get them away.' The camera turns to a man in the next room and we hear him reply, 'What? What's the problem?'

Here the camera was used without any intention of capturing an idealized representation of home life. In the hands of a teenager the camera becomes a tool for rebellious and transgressive behaviour. Mother is humiliated and father is shown as uncaring or inattentive. The daughter implicitly consents to the filming, and the event appears to be framed by the intention to put the moment up on YouTube. This aim is reinforced by the daughter's description of the video on YouTube: 'Yeah, my mom's drunk and on the toilet. Stupid bitch. Haha.' This is not *The Adventures of Ozzie and Harriet.*

As harsh as this video is, it demonstrates that Moran was correct when he asserted that 'video and television share many of the same codes to represent everyday family life.'[19] Print advertisements for products such as Candy (women's underwear) show women on the toilet. The toilet is a common fixture in humorous television advertisements (mostly European) and in many genres of movies. There are thousands of amateur videos on YouTube that depict people using the toilet, throwing up in the toilet, and even sleeping on the toilet. The proliferation of scenes in movies and television that depict the use of the toilet must surely be an influence upon amateur videographers. One suspects that bathroom scenes were not as prevalent in home movies as they are among the new generation of videographers.

The particular choices that any one YouTube video represents are the result of a complex set of factors. The economic and technological character of video makes it much more likely that children will use the camera and that their video practices will capture the most banal situations and scenes. At one point in the second drunk-mother video the videographer pauses to film a bottle of glue. This in turn leads his friend to exclaim, 'Why are you filming glue?' No matter how inappropriate the occasion or how trivial the subject, the camera remains on and recording. His friend's intoxicated mother sitting on the toilet or a bottle of glue – both are treated as appropriate subject matter by the young videographer.

Technological and economic factors do not provide us with a complete explanation of why young videographers are choosing to represent the home and family in the way that they do. Video practices are also influenced by current conventions of representation in commercial media, which provide amateur videographers with standard scripts, characters, and a repertoire of possible scenes and scenarios. Thus, the reproduction of the home in domestic videos is influenced by how it is represented in commercial media.[20] The bathroom and the toilet have been normalized as a setting within advertisements, television, and cinema; consequently, their appearance in home videos is not an aberration but a confirmation of a larger pattern of domestic representation. Such background contexts of representational patterns may or may not influence the making of any particular YouTube video. We do not know to what extent the representation of the toilet in popular culture has influenced the choices and intentions behind the drunk mother video. We do know, however, that the toilet is part of conventional representational practices in print, television, and film.

In another home video on YouTube (348 views) we encounter a videographer who uses the camcorder to confront, torment, and humiliate an intoxicated, elderly woman. The scene is the woman's bedroom. Next to her bed sits a small fridge. She is seen in her underwear clutching a two-gallon bottle of red wine (nearly empty). For almost eight minutes we witness the videographer, apparently her middle-aged son, rage at the woman and reduce her to tears. Old grievances are discussed. The videographer accuses the woman of 'pill popping,' drunkenness, and neglect. She responds that she is the one who is neglected. The family dog occasionally rises from the bed and licks the woman's tears away. The scene is altogether heart-rending. One viewer comments, 'why would anyone put anything like this on YouTube.' YouTube is home to America's saddest home videos.

The altogether shocking way in which family members discuss each other and record each other on YouTube is not unprecedented in media culture. People are constantly seen disparaging each other on reality television and talk shows. Parents and their children regularly appear on *Dr. Phil*, the *Oprah Winfrey Show*, and the *Jerry Springer Show* and reveal the most intimate details of their lives. This genre of television promotes a context of narcissistic self-revelation (that seldom involves only one-self!).[21] Television's portrayal of confession and confrontation provides the cultural backdrop to the practices of amateur home videos. Confrontation becomes an occasion for using the camera to record action, post a video to YouTube, and gain an audience.

Technology, economics, the surrounding media environment, and the awareness of YouTube as a distribution platform are some of the factors that shape domestic video practices. Another aspect of the media environment that influences the amateur videographer is the desire for attention. YouTube provides children and teenagers with a reason to turn the camera on their siblings and parents. Youths post videos online as a way to expand their social network and increase their self-esteem by garnering viewer hits. Of course, there are many other reasons, but one of the most prevalent is the desire for subscribers. A search for videos with the phrase 'please subscribe' in their titles or descriptions delivers hundreds of thousands of hits.

All YouTube videos made by children are not as grim as those featuring drunken parents. Many, such as *13 Reasons I Love My Mom*, depict children extolling their parents' virtues. Mother's Day has become an occasion for posting videos such as *To My Mom with love Happy Mother's Day* and *Mother's Day Video 2007*. The commentary section of these videos often becomes a gathering space where siblings, parents, and relatives exchange thanks and greetings.

The video *Happy New Year 2008* depicts a Canadian family sending a digital 'shout out' to their friends and distant relatives as they stand in the snow outside their house in British Columbia waving sparklers, playing *Auld Lang Syne* on kazoos, and counting down to the new year. Moments of ritualized celebration – Mother's Day and Father's Day, birthdays, family reunions, wedding anniversaries, and even deaths – are posted to YouTube. Individuals use amateur video as a way to celebrate their families and friends, commemorate special occasions, and stay in touch with one another. Video greetings, slideshows, home videos, and vacation videos are exchanged over YouTube as new forms of digital greeting cards. Through these media practices YouTube enhances social connections across distances and represents a virtual space where people come together to create and maintain social ties.[22]

Siblings on the 'Tube

When teenagers are not filming their parents, they are busy filming each other. Siblings can often be seen tormenting and playing pranks on each other on YouTube. One video depicts an eighteen-year-old girl being grounded after going out to meet a boy she encountered on MySpace.[23] When she returns home, her teenage brother begins filming from a hidden location and quickly inserts himself into the middle of the conversation between his mother and his sister. After a few moments the camera's presence is known to both mother and sister. From behind his camera the brother taunts and torments his sister for three minutes. Throughout, the mother repeatedly tells her son to stop filming. Finally, she insists that he delete the video and turn the camera off (obviously, he ignored the first demand). He tells his mother that he intends to put the video on YouTube and laughs after he has reduced his sister to tears. He brags, 'this is a family video that will go down in history.' The video was viewed over 2,500,000 times and garnered more than 12,000 written comments. The majority of the comments congratulate the boy for humiliating his sister.

This video has been posted to many sites across the Internet and the videographer has boasted about the number of views it receives. This disturbing and harmful form of media practice is now common. Increasingly, video cameras are used to capture private and embarrassing moments of friends and family members and then post these incidents to YouTube. Many parents and teenagers fail to understand the possible lifelong consequences of such actions.

Moran suggests that home video is a 'less self-conscious presence on the scene of domestic living, [its transparency] tends to relax some of the artificial conventions imposed by home movies.' Two factors, in particular, are responsible for making the use of video cameras a less self-conscious affair. Video cameras have proliferated across the social world. Many cellular phones also function as video cameras. Also, most forms of domestic video cameras can be used without concern for the cost of film and processing (which are non-existent). Thus, where the use of the more expensive home movie camera led to the selection of 'only a narrow, partial spectrum of everyday life,' which in turn led to the representation of 'an idealized image of home,' the much less expensive home video camera encourages its user to capture more of the ordinary and more of the extraordinary action of daily life.[24] The video camera is a constant presence in the lives of the YouTube generation, and this may render its use more transparent within the family setting.

Children and adolescents are among the most prolific contributors to YouTube, a pattern that holds across other forms of Internet content creation as well. Over 50 per cent of blogs are written by children and youth under nineteen years.[25] A BBC report notes the phenomenon of seven-year-old bloggers, and we find that children are creating Internet content at increasingly younger ages.[26] Children are now being taught in school how to contribute material to a blog, and a recent television advertisement featured a small child demonstrating how easy it is to upload pictures from her camera to the Internet. A new generation is being socialized to act as Internet content producers when very young.

Daniel J. Solove's study of reputation, gossip, and rumour on the Internet describes how we are collectively engaged in a massive experiment in the construction of identity when we post so much of our lives to the Internet.[27] The YouTube generation is blogging about their every thought and action, using Twitter to post instant updates on their current activity; revealing personal data such as telephone numbers, addresses, and birthdates on Facebook and MySpace; and capturing mundane, embarrassing, and downright damning moments of each other's lives on video. Children at one school were found to have revealed birthdates (88 per cent), phone numbers (40 per cent), and home addresses (51 per cent) on Facebook.[28] The YouTube generation is revealing online an astonishing amount of detailed personal information about themselves, their families, and their friends. The examples of the drunk mothers and the disciplined daughter explored earlier align with Solove's suggestion that the YouTube generation 'may find it increasingly difficult to have a fresh start, a second chance, or a clean slate.'[29]

Children do not realize that, in shaming their parents or their siblings online, they are also damaging their own reputations. Armed with video cameras, they are transferring private family affairs into the public realm and in so doing are transforming a passing moment into a permanent and widespread public memory. Our careless teenager risks permanently redefining his sister's identity (she is now known to hundreds of thousands of Internet users by a derogatory nickname). He will never be able to erase the video from the Internet and risks being labelled as cruel and mean. He risks his future career in the new job environment where human resource personnel regularly use Internet applications such as YouTube, Facebook, MySpace, and Google to learn more about their potential hires. University admissions boards are also conducting online background checks and rejecting applicants based upon what they find on blogs and social networking sites.[30] Careless and unethical video practices risk damaging a person's ability to make new friends; many students use the Internet to do background checks on individuals they are considering dating.

In the past, home movies were typically viewed a few times and then relegated to the basement or attic. Home videos on YouTube may be replayed many times, which suggests that they will be more consequential for personality development than home movies. Repeated viewing of home videos may turn out to be similar to repeated exposure to advertising among children. It could reinforce the impressions made by the video and increase its degree of persuasiveness.[31] Digital home videos are destined to play a much larger and sometimes much more nefarious role in the construction of our identities than analogue home movies. Fortunately, domestic video practices also capture moments of affection and love. Yet even innocent action can been misinterpreted when viewed at a distance. Videos of brothers and sisters playing together in their bedrooms are frequently subjected to lewd comments.

Adolescents are expressing their sexuality through amateur home video. Videos made by teenagers and tweens explore issues such as masturbation, sex, incest, and rape. Teenagers and children are seen engaged in lewd dancing to highly graphic songs. They often use words such as rape, sex, and incest in their video titles even when the videos have no sexual content (this is done simply to attract attention to their videos and their channel). There is a large body of videos that show teenagers and children at home cross-dressing. Home movies of the past tended to exclude such representations of youthful sexuality.

Even under normal circumstances children exhibit a broad range of sexual behaviours.[32] Sexual play in children and teenagers is a natural and healthy process, but its presentation on video is filled with risks. YouTube is playing an ever larger role in representing this normally hidden aspect of domestic life. Floyd Mansfield Martinson notes that even advanced technological societies such as ours go to 'great pains to restrict sexual activity among children.'[33] The taboos that surround the subject are bound to collide with the increased exposure to sexualized play that YouTube is creating.

Brothers and sisters also film each other playing musical instruments and video games, dancing, and engaged in a wide variety of other activities. It is probably the case that being filmed by a sibling while playing (or doing anything at all) is increasingly experienced as a normal part of the life world of children and teenagers. Children grow up with the ubiquitous video camcorder in the hands of parents and relatives; for them, it is part of the background environment of practices and attitudes that constitute taken-for-granted experiences, as is seen in the often heard comment, 'put it on YouTube.' As a result, we are entering a new era wherein children are creating a substantial volume of media content and drawing significant global audiences within the Internet.

A teenager from Nebraska, Lucas Cruikshank, produced YouTube's first popular weekly series to reach over 1 million subscribers. When he was fourteen, Lucas created the character Fred Figglehorn, a lonely six-year-old with anger management issues and a recovering alcoholic mother. Fred draws as many as 27 million viewers to each week's episode and garners tens of thousands of comments. Fred is one of the rare examples of highly successful programming for children by children. Cruikshank's fame has won him endorsement deals from Hollywood studios and media coverage from sources such as CNN, the *Los Angeles Times*, *BusinessWeek*, the *Wall Street Journal*, and the *Sydney Morning Herald*. Fred became the fastest growing channel in YouTube's history. Cruikshank has signed with a talent agency and is now selling product placements on his show. Education researchers are looking into Fred as a potential educational resource and for insight into the hyperactive YouTube generation that Fred represents.[34] This is part of a larger trend that sees educators looking to YouTube for pedagogical content.[35]

Swearing Babies, Laughing Babies, Biting Babies

Whereas the first moving images of a new family member were once recorded by a parent, they are now just as likely to be recorded by a young sibling. When parents record their children's actions on YouTube, they often show judgment that is no better than their children's. What is cute and innocent behaviour in a private context may be subject to misinterpretation in the public sphere and become an indelible part of a person's identity. Consider how swearing babies are generating audiences measured in the millions of views on YouTube. These moments are part of every parent's experience. They are perfectly normal aspects of childhood development, but one wonders about the long-term result of so many people having such moments of their lives recorded on the Internet.

Tens of thousands of baby videos populate YouTube. By far the most watched videos in the family genre are those of laughing babies. The most popular video in the baby subgenre is BlackOleg's clip titled *Hahaha*. Under two minutes long, this video depicts a baby in a highchair laughing infectiously as his father occasionally says 'boo!' The video was viewed over 100 million times and generated 145 video responses and 241,584 text comments. That a laughing baby can become the tenth most popular YouTube video is quite extraordinary. Unlike responses to videos depicting cruelty towards siblings, the comments found among laughing baby videos are overwhelmingly positive.

Another of the most popular YouTube videos features a baby named Charlie and his older brother Harry. *Charlie bit my finger – again!* provides us with an opportunity to explore how the use of YouTube as a family photo album and screening room complicates the construction and maintenance of personal identity. This video has been viewed over 130 million times and generated over 329,051 comments. *Charlie bit my finger – again!* depicts a baby sitting in the lap of his three-year-old brother, Harry. The older sibling laughs at how his brother, Charlie, bit his finger. Harry then once again puts one of his fingers in Charlie's mouth, but this time Charlie bites down quite a bit harder. We witness Harry's initial delight turn to concern and then pain: 'Charlie, that really hurt!' Charlie laughs and Harry's frown turns to a smile as well. Less than a minute long, the video captures a delightful moment of play and discovery. The father offered the following explanation of the sequence: 'Even had I thought of trying to get my boys to do this, I probably couldn't have, neither were coerced into any of this and neither were hurt (for very long anyway). This was just one of those moments where I had the video camera out because the boys were being fun, and they provided something really very funny.'[36]

Charlie bit my finger – again! is an important moment in amateur online video, as it provides us with an opportunity to explore the effect of a

YouTube video on a family over time. The first video clip of Charlie and Harry appeared on YouTube on 22 May 2007. Since then, the parents have put more videos of the boys on YouTube. Brief clips show the two brothers growing up together. A more recent family video shows Harry and Charlie with their new brother, Jasper. The vast majority of comments are positive and affirm the boys as loving and adorable. None of the other videos has attained the same level of fame as the first one, but the viewing numbers indicate that many people continue to follow the lives of Charlie and Harry.

Among the family's collection of online home videos we find the parents, Howard and Shelly, along with the boys, giving their first television interview. Through the television interview we learn that Howard put the initial video on YouTube because the boys' godparents and other family friends live overseas. The interviewers talk directly with Harry about the initial finger-biting incident and affirm that Harry was a good boy (as he forgave his brother for biting him) and they reassure Harry that he 'loves his brother, don't you? Lots and lots' (*Charlie Bit Me – The First TV Interview – Richard and Judy*). Harry nods his head in agreement.

A year after first uploading the video to YouTube, Howard provided the following insightful reflection: 'I have discovered over the last year whilst moderating comments, YouTube is a forum for opinion not debate. I mean by this that there is little point in trying to reason what someone has said and reply. If someone is rude, ignore it, the problem lies with them. If it is offensive I delete it. Negative tirades are nearly always uneducated hate or resentment, at worst anti-social or anarchic. It is a shame that opinion is becoming less empathetic every day. It is easier to hate than respect.'[37] His experience captures the darker side of YouTube culture – fame attracts hatred.

Initially the subjects of an amateur home video made by their father, Charlie and Harry are now media personalities. They have been on British television and continue to draw audiences with new video clips of their lives. They have also been widely parodied by other amateur videographers. Dozens of remixes, such as *Charlie Bit Me Techno Remix*, turned the clip into music videos and have garnered over 1 million viewers. The comments that the music remixes gather tend to be much cruder, which suggests a slightly different audience. YouTubers have also made original songs based on the video (such as *Ainslie Henderson – Charlie Bit Me Song*). The scenario has been parodied with vampires biting someone's finger (*Dracula bit my finger and it really hurt*). It has been re-enacted by children (*Charlie Bit Me My Finger – 5 years later*), teenage boys and girls

(*Charlie Bit Me 10 years later*), young adults (*Charlie bit me 15 years later*), and middle-aged men (*Charlie bit me 50yrs later*). Each one of these scenarios has numerous variations on age, gender, race, and language. Over 2,000 remixes and parodies have been made using appropriated material from television shows, animations, music videos, and celebrity pictures.[38] Teddy bears and family pets have been used as stand-ins for Charlie and Harry.

Now that YouTube acts as a global family room for the screening of home videos, we are faced with a host of issues, some known and many as yet unknown. We do not yet know what the consequences will be for those who grow up not merely watching YouTube but being watched on YouTube. We are taking formerly private areas of our lives, recording them, and then stripping them of their context by broadcasting them across the Internet. The lack of context can make it difficult to understand the action in a home video. Innocent behaviour can be seen as deviant. We are globalizing the context of our selfhood. We are transforming passing moments of our childhood into moments permanently fixed in global media culture.

The opinion of others plays a role in the construction of our identity and the ongoing maintenance of our self-esteem. Our sense of self is strongly dependent upon what we think others think about us.[39] But not all contexts are equal when it comes to forming our self-concept. What we do in public can be more consequential to our self-identity than what we do in private.[40] Public evaluations of the self can be more consequential than private evaluations of the self.[41] By moving more of our behaviour out of the private domain and into the public domain, amateur video practices make larger tracts of our lives consequential for our sense of ourselves.

Psychologist Dianne M. Tice explores how actions performed in public affect us differently from actions performed in private and confirms that 'public behaviour implicates the self more than private behaviour.' There is considerable evidence to support her conclusion that 'other people's perceptions constitute an important part of the self.'[42] Originally proposed by C.F. Cooley in 1902, this theory of the 'looking glass self' has withstood over a century of scrutiny.[43] YouTube is a new form of looking glass; it enables others to apply deviant labels to anyone, which may lead individuals to internalize those labels. Such attitudinal changes in a person's self-concept may be temporary or permanent. It is not yet certain if the looking-glass effect occurs through the Internet's mediated social environments.[44] Will the wired generation learn to see themselves as the YouTube audience do?

Normally, self-identity is constructed partially through face-to-face relationships with a relatively small number of individuals. Through YouTube we are dramatically increasing the number of individuals who can influence our self-perception and enabling them to manipulate our intimate memories. We are allowing a group of strangers to tell us what they think of us. Psychologists describe the definition of the self as a relational process that depends upon whom we interact with and the context of interaction.[45] YouTube dramatically alters both coordinates. We allow strangers to tell us who they think we are, and those strangers on YouTube are often vicious or obscene.

Amateur home videos complicate the processes of self-definition and bring in their wake great risks. Consider the case of Charlie and Harry. Their childhood memories are now the common property of the Internet community. These memories in and of themselves are happy and healthy. Yet now they are subject to appropriation, remixing, and manipulation. Harry and Charlie may be repeatedly reminded of their biting video for the rest of their lives, so the way the video is received and appropriated will be important to their sense of self. They face a future that is partially defined by events of their childhood that have been posted to YouTube, appropriated, and reinterpreted.

We may be creating conditions for self-construction where the past plays a much more current role in our lives. In Charlie and Harry's case, this may prove to be benign. In the case of children shaming their siblings or their parents by recording transgressions, the damage may be long lasting, the memories painful and indelible. When teenagers inappropriately use the video camera to make public a minor misdeed, they violate the private space that the home affords to individuals. This privacy serves an important function by protecting 'minor non-compliance with social norms.'[46]

Adolescents armed with video cameras deprive family members of the right to determine autonomously what information they will reveal to others, and also rob them of the privacy of the home, where they can indulge in minor acts of non-compliance with social norms (burping, farting, swearing, cross-dressing, picking one's nose, and so forth). Both psychological studies and introspection tell us that we behave in one fashion when in public and in quite another fashion when in private. Privacy works by creating spaces for individual autonomy. Amateur home video practices can disrupt the individual's autonomy and may do so in the important early years of identity and personality construction.

In Charlie and Harry's case there is a toxic parody, *Charlie bit my finger – Adult remix*, which transforms the original video into pornography.

This parody represents the two boys as young adults engaged in fellatio; it uses the original voice of Harry and has garnered over 1 million views and thousands of comments.

BlackOleg's *Hahaha* video has also been appropriated and turned into various versions of a laughing evil baby. In appropriations such as *Mean Laughing Baby/Fall Compilation* (2 million views) we see the laughing baby mixed in with a montage of clips that depict people involved in accidents. Childhood memories are subject to both mild and extreme forms of appropriation and both forms proliferate within YouTube. When parents use YouTube as a global screening room for home videos they open their children's past to drastic forms of reinterpretation. YouTube is changing how family memories are preserved and may be altering how identities are constructed in the public sphere.

Solove describes the loss of privacy on the Internet in terms of the permanent alteration or destruction of a person's identity: 'Information that was once scattered, forgettable, and localized is becoming permanent and searchable. Ironically, the free flow of information threatens to undermine our freedom in the future.'[47] When children and teenagers make poor ethical decisions regarding how to represent themselves and others, they limit their options in the future and restrict their ability to reinvent themselves without being unduly chained to past actions and memories. To varying degrees, information is not only permanent and searchable within the Web, it is also subject to appropriation and radical alteration. In the first major exploration of the psychology of the Internet (1999) Patricia Wallace argues that 'we have far more control over our self-presentation on the Internet compared to what we had in high school.'[48] This conclusion needs to be revised in light of YouTube. It is hard to see how we have any significant control over what others do with our information and self-presentation on the Internet in whatever form it appears.

Children and adolescents are among the earliest and most active adopters of new media.[49] YouTube is the favourite online destination for children two to eleven years old. Children under eighteen view more YouTube clips than any other age group.[50] Any way you slice it, children are becoming deeply entwined in YouTube and they are a significant part of the online audience and produce a significant amount of online content.

Amateur home video includes practices such as children broadcasting live video feeds to other children and adults. Webcams and live Internet feeds are also used by children for visiting with distant relatives, which may lead to new experiences of 'liveness' (real-time co-presence) in so-cial relations.[51] Children and adolescents are among the heaviest users of live webcams, which often are located in bedrooms. The *New York Times* describes how teenagers make use of one live chat site, Stickam.com, in the following terms: 'several thousand of its mostly teenage members log onto the site each night to broadcast their own lives, often from their bedrooms. They put on makeshift talk shows, flirt with other members in video chat rooms, and often, if they are female, field repeated requests to take off their clothes.'[52] Stickam's parent company was reported to have close ties with the pornography market.

The practices of home moviemaking were once defined by adults. Now the leaders in home video may well be children and adolescents. This possibility aligns with Sonia Livingstone's observation that chil-dren and young people often 'lead the way' in the use of new media.[53] YouTube demonstrates that, in the Internet age, the adoption of video technology by children and adolescents has eroded the authority of par-ents over home media practices, just as the state and the market have lost hegemony over media consumption and production in the Internet age. Within the home of the Victorian bourgeois family the relationship between children and media consumption was very strictly controlled.[54] The era of such control has long passed. Even a decade before the mass adoption of the Internet Joshua Meyrowitz described how parents had little jurisdiction over the images and ideas their children were consum-ing.[55] YouTube is exacerbating this loss. Poster's description of the me-dia environment of the contemporary home is fitting: 'The home has become infinitely permeable to the outside world.'[56]

We know little about the impact of new media technologies on child development.[57] The bulk of studies on family and the media focus on children and adolescents as the audience of mass media. What we have not yet faced is the emerging role of mass numbers of youth as producers of huge volumes of content. We have never before seen amateur home videos function as a major source of entertainment for children *and* as a source that features children as content in quite the same way that is now occurring online. What will the effect of this be? Even after the presence of television in our lives for over half a century, some research-ers can only conclude, 'The real import of children living and grow-ing in a media-saturated environment is only rudimentarily understood. We know far too little about the role that media play in developmen-tal outcomes.'[58] Other researchers have come to very different conclu-sions. A meta-analysis of 173 studies found that 80 per cent agree that heavy media exposure increases risks in the areas of obesity, smoking, sex, drugs, alcohol, and learning.[59] Health and sexuality researchers are most certainly correct when they conclude that the media act as 'super peers,' highly influential friends in teenagers' lives.[60] Without careful adult guidance and surveillance YouTube will not be a good friend for our children.

Summary

Moran suggests that we define the specificity of video, its essence, according to its properties that lend themselves to 'specific social functions.'[61] In other words, the essence of video is determined not solely by its technology but by how its technology creates social uses for video. One cannot help but be struck by how video is being used to shame friends and family members, attack reputations, and alter childhood memories. Shaming people and negatively appropriating family members is not the essence of home video, but it is a prominent social function of video. Home video is also used to praise siblings and showcase their talents, express love and affection between family members, and reach out across distances, so we must be careful of overstatements. We see a vast spectrum of domestic behaviour on YouTube, including sentimental representations of domesticity, the most brutal moments of sibling rivalry, and sexualized play of children and adolescents.

Home movie practices of the mid-twentieth century once reflected a dominant ideology and a coherent world view of the middle and upper classes, and they communicated stability and conformity to norms. Home movies presented a much narrower slice of domesticity. Moran's analysis of the points of continuity and discontinuity between home video and home movies concludes that home video 'reveals that families have always been more complex and contradictory than home movies have generally portrayed them.'[62] Digital camcorders allow amateur moviemakers to explore random moments, use more outdoor locations, and shoot scenes without concern for organizing the participants into posed, conventional arrangements.

The low cost of actually using a video camera along with its much extended recording time and better performance in low lighting 'substantially increased the potential range and volume of events and behaviors' that could be recorded.[63] Larger areas of daily life, more hidden areas of home life, and more trivial and mundane aspects of the everyday are now regularly caught on camera and posted to YouTube. YouTube is the world's living room. This virtual living room invades our privacy, erodes our autonomy, and threatens essential social dynamics such as the need for moments of private non-compliance.

What is fairly certain is that the use of expensive domestic film cameras was primarily the domain of adults and represented the home and the world through adult eyes. As yet we have no numbers to assess the degree of shift that has occurred, but it is clear that there is a growing army of young people who are armed with video cameras, YouTube accounts, and plenty of attitude. The convergence of the Internet with the mass deployment of computers, consumer electronics, and digital cameras is changing how the family is represented. It may also be changing our children's identity and their future prospects.

3 Video Diaries: The Real You in YouTube

In June 2006 a young girl named Bree made her first video blog and spoke about how she likes a lot of the people on YouTube. She told viewers that her town is really boring and nervously reflected that she felt she was a 'dork.' *First Blog / Dorkiness Prevails* was the YouTube community's introduction to the video diarist Bree also known as Lonelygirl15. Bree and video blogging on YouTube almost immediately became a media sensation. Within four months of her first YouTube appearance it was discovered that Bree was actually Jessica Rose, a professional actress.[1] Yet for a short while Bree was the poster child for YouTube's community of amateur video diarists.

Sean Cubitt describes the video audience as having an acquired set of skills 'so deeply embedded we scarcely know we have them and rarely stop to value them.'[2] These skills allow the audience to distinguish between newscasts, dramas, soaps, situation comedies, and advertisements or follow complex styles of storytelling. Writing in 1991, Cubitt made the intriguing suggestion that video improves the audience's media competence and interpretation skills. We see this when the YouTube audience invariably exposes commercial productions that attempt to pass themselves off as amateur productions, as in the case of Lonelygirl15.

Video diaries provide an occasion for exploring the YouTube audience's demand for high levels of authenticity from fellow YouTubers. Michael Wesch suggests, 'If you could name a core value on YouTube, it's authenticity. The strongest critique is to say that you're hiding behind something or you're not being real.'[3] The authenticity and rawness of amateur video diaries have been described by John Corner as a form of 'punk television.'[4] Punk rock is characterized by a lack of technical sophistication and acts as a challenge to the overproduced 'fakeness' of more popular rock bands. In a similar way video diarists forgo sophisticated forms of storytelling and production so as to be more real. As in the case of punk rock, a segment of the audience is drawn towards the raw and more genuine quality of video diaries. Punk rock undermined rock and roll's claim to be the real music of rebellion. Similarly, video diaries undermine television's claims to authenticity and the real. Video diaries and the growth of camcorder culture (the widespread use of video cameras) have helped to demystify the techniques of television's production of reality. This in turn, suggests Corner, has undermined the attraction and the power of mainstream broadcast television. Slick editing, lighting, and flawless dramatic performance gave Bree away as a fraud. Lonelygirl15 was too real to be real.

Claims that YouTube offers a more authentic experience abound on the Internet. Although these claims raise a host of largely unsolved epistemological issues about the nature of the 'real,' they nonetheless represent significant social facts. Mara Fleishman, a producer of a New York cooking show, sees online video as 'definitely more authentic than television.'[5] Fleishman's position as a commercial video producer obviously taints her claim regarding video's authenticity, but this need not muddy the waters. Tourism British Columbia is using amateur videos that capture tourism experiences throughout the province, because 'nothing is more credible than real experiences from real people.'[6] Blogger Bob Jacobson suggests: 'The appeal of amateur fare isn't that it's somehow less demanding than big media and therefore easier to view and download, to take away (as in fast food). The appeal is that it's authentic and about people. There's nothing more interesting to real people (not Hollywood producers) than authentic stories told about other real people.'[7] YouTube's rapid transformation into a mass medium is partially explained by the perception that amateur video offers something that television does not. That something is often described as more real.

The Real Me

There are hundreds of videos on YouTube titled *The Real Me* that explore the difference between videographers' on-screen persona and their 'real' self. College student and comedian Kevin Wu's YouTube channel kevJumba features a video called *The Real Me* (511,359 views) wherein he explains how he is a different person from his videos, 'You don't know me … You don't really know someone through the Internet.' Here we see a performance artist dealing with the difference between the audi-

We can never be certain that an authentic self is captured in an amateur video.

ence's reception of his character and his offline personality. This type of conversation between YouTube video bloggers and their audiences is an emerging subgenre. Another YouTube comedian, Michael Buckley, also has a video titled *The Real Me* that addresses the same issue – who really is Michael Buckley? Dozens of amateur video diarists respond to Buckley's invitation to address the same issue.

Gareth Smith replies in his video *Re: The Real Me*, 'I am the real me in my videos but there is a boundary where I am not.' Common among many of the replies to Buckley is the reflection that there is not an exact correspondence between a video and the everyday self. Fourteen-year-old Jimmy's response provides insight into a particularly toxic feature of YouTube when he admits, 'I act like more of an asshole on YouTube ... because it is easy to say shit behind a computer.' Likewise, the teenage boy behind THEFRANKSHOW YouTube channel relates, 'I am not as nasty in real life.' When this young boy is in front of the Web camera he feels 'powerful,': 'like I can express what I feel ... it just really gives me that extra boost to be nasty and mean like "ah ah I'm in front of the camera and you cannot hurt me."' Significantly, the boy feels less powerful when text blogging and more powerful when video blogging. Another male video diarist also admits, 'I tend to be a jerk, very opinionated, arrogant, just kinda an ass, and it's really not the case [that I am like this] in real life ... I never fully understood why I let loose here on webcam.'[8] It may be the case that this type of aggressive or negative transformation of personality within online diaries is particularly prominent among young males.

Regardless of the jackasses and haters that populate YouTube, its members do experience a sense of connection. One young female video diarist confides that she finds 'more understanding people on YouTube than she has in real life.'[9] The creator of the NannyCam YouTube channel, amateur video diarist Shel, replied to Buckley's challenge. Shel confided, 'definitely what I show you on YouTube is a part of me, but there is much more. I do not tend to discuss the negative stuff ... maybe there's stuff that I do not want to admit to myself.' Shel explores her childhood and family relationships and reveals yet more layers of herself. Even in the course of confessing that they do not fully reveal themselves video diarists delve deeper into their biographies. One viewer responds, 'I love finding out about the people behind the videos what they are actually thinking inside their minds, and when they share it I feel like I get to know them a bit more.' Curiously, Shel records her video in the privacy of her car and at one point is driving and talking to the camera on her dashboard. Here we see how varied the locations are for recording video diaries.

Video diarist Lauren contemplates the long time it takes to get dressed up, put on make-up, and edit out material that she thinks is stupid. Lauren takes off her make-up on camera and gives the audience an invitation: 'meet the real me ... if you were looking for perfection turn on your TV. This is YouTube. This is real and this is who I am.' Another young woman confides her omissions: 'all the things that I edit out of my videos, all the little pieces of me that I don't want you guys to see because I am afraid of what you guys are going to think.'[10] 'YouTube is all about being yourself ... but sometimes I feel like instead of sharing myself with the world I am hiding behind the camera and only letting people see the parts of myself that I want them to see ... This is my channel. This is my life … I want you and everyone to know that I am real, not just a face, not just a voice,' declares another young woman.[11] These confessional videos give the impression that online diarists often feel as though they should be fully authentic and transparent for their audience, but that for one reason or another they fail to meet this ideal.

There is a contradiction in contemporary consumer society that pits the hyperreal, fake, and plastic against the authentic. Karen Wright describes this contradiction in the pop-psychology magazine, *Psychology Today*: 'increasingly, contemporary culture seems to mock the very idea that there is anything solid and true about the self. Cosmetic surgery, psychopharmaceuticals, and perpetual makeovers favor a mutable ideal over the genuine article. MySpace profiles and tell-all blogs carry the whiff of wishful identity. Steroids, stimulants, and doping transform athletic and academic performance. Fabricated memoirs become best-sellers. Speed-dating discounts sincerity. Amid a clutter of counterfeits, the core self is struggling to assert itself.'[12] As consumers and audience members we long for both fantasy and authenticity. Perhaps the massive outpouring of self-reflection and video confessions on the Internet is an indication of a contemporary crisis of the real, the self, and the authentic.

The current flood of self-disclosure in blogs and video diaries may be a response to the fragmentation and uncertainty that are said to characterize the present postmodern moment. Mary Evans suggests that confessional discourse offers people 'a chance to stabilise the uncertainties of existence.'[13] The twentieth century has been the high age of uncertainty. The historian Eric J. Hobsbawm saw this time as defined by a 'hunger for a secure identity.'[14] Empire itself (or at least its centre), the American nation, is widely thought to be in the throes of an identity crisis. Evans sees confessional discourse as something that allows people to 'convey the impression that their lives are lived in orderly and coherent ways.'[15] Autobiography reflects a deep desire for order and stability. Ironically, the preferred medium for self-expression and identity construction, the Internet, is the very medium that destabilizes identity.

Digital Diaries and the Confessional 'Tube

YouTube and the Internet are not the first instances of a mass outpouring of confessional discourse. The Victorian age was also intensely autobiographical. Confessional practices have changed throughout history, and the form and meaning of diaries continue to change in this digital age. The definition of autobiography and diary is as murky as the definition of video itself, so when we locate these practices in a digital environment, their precise definition becomes that much more elusive. For example, writing in 1930, Donald Stauffer suggested that diaries have 'scant claim to consideration,' and he completely dismissed the possibility that a diary could attain the status of a work of art.[16] Such high-culture pretences are no longer convincing, yet there is little consensus on exactly what constitutes autobiographies or diaries.

More recently, George P. Landow has proposed that autobiography designates only narratives that 'self-consciously contrast two selves, the writing "I" and the one located (or created) in the past.' Landow proposes an essentialist notion of autobiography, one that admits into its fold only a very narrow spectrum of subjectivity. Yet there is little justification for defining the genre of the autobiographical by privileging a particular relationship to the past. Landow is concerned with preserving autobiography as a 'clearly defined literary mode' at a time when all genres defy clearly defined borders.[17] This is, at best, a rearguard action against an overwhelming tide of change. We can no longer define autobiographies and diaries as having a particular subjectivity, theme, pattern, structure, or meaning, as their forms are far too diverse. Herein I treat any YouTube video that has some confessional or self-representational quality as belonging to the autobiographical and diary genre. Digital diaries cross boundaries of genre and media practice. The YouTube videos of a performance artist such as Chris Crocker can have a confessional quality that lends itself to the diarist genre.

The self is both represented in the diary form and constructed through it. We cannot claim that the diary is a simple representation of the self, but neither should we reduce the self to the point where it is a mere by-product of its contexts and mediums. The diary is the location of unstable, contested, multiple, and often incoherent selves, but it is also a place where we encounter the real of others. The online diary form may be flawed, but it can provide us with a representation of social reality.

Confessional videos on YouTube represent one of the most dramatic changes in a diary form of the past, which was mostly a private, secret practice. Diaries once were a manifestation of an increased demand for autonomy and privacy. A new type of individual emerged in the seventeenth century and a new relationship to the private sphere developed. Felicity A. Nussbaum writes that the diary 'signifies a consciousness that requires psychic privacy in a particular way.'[18] Could it be that the transition of private diary practices to the public domain of Web blogs and YouTube videos indicates yet another shift in the nature of consciousness? Perhaps the contemporary self requires psychic publicness in a particular way.

Nussbaum provides insight into another aspect of the movement of diaries into the public sphere when she suggests that the 'marginalized and unauthorized discourse in diary holds the power to disrupt authorized versions of experience.'[19] Online diaries stand alongside many other forms of Internet-based representational practices that disrupt authorized versions of reality. Throughout YouTube, individuals are seen challenging normative notions of what it is to be gay, black, male, female, and so forth.

Are online diaries a new media practice? Jay David Bolter suggests that there is a 'widespread assumption that the principle task of media producers and critics is to identify the elements that make new media truly new.'[20] There are plenty of voices that remind us of the fallacy of essentialism, yet we need not shy away from insisting that there are indeed certain unique aspects of the Internet. The mass production of online amateur videos is similar to but also different from the representational features of earlier technologies. The point is not that online diaries are utterly new, as they are clearly a digital version of an ancient practice. Yet the combination of features such as global distribution, mass involvement (as diarists and audiences), malleability, and audience interaction within online diaries is unprecedented.

Bolter claims social computing is 'self-referential, fragmented, and multiple – the antithesis of the aesthetic of transparency.'[21] Yet are the Internet and social networking sites like YouTube truly the antithesis of transparency? Claims about the Internet's being the antithesis of transparency ring hollow when two people can connect via a live video stream and have a conversation about transparency. Transparency must certainly be an adequate description of *some* aspects of video diaries. Online diaries are complex, contradictory, opaque, and transparent. So, while terms such as *real* and *authentic* are as problematic as 2,300 years of philosophy, we gain little by explaining away the more commonly felt experiences of the YouTube generation. Bolter and Richard Grusin's original objection to the claim that cyberspace will produce a virtual realm that is more real than real was an important correction to certain postmodern excesses in new media theory.[22] Yet we need not throw out the real baby with the virtual bathwater when trying to define the quality of the real in online video.

A search for the term vlog (video blog) on YouTube renders 310,000 hits – an imprecise metric but one that nonetheless speaks of a certain attention economy within the Internet.[23] We are engulfed in an overabundance of websites, channels, voices, videos, and blogs, all demanding our attention. The confessional spaces of the Internet only add to the oversupply. As written diaries once did, video diaries continue to provide us with valuable insights into the lives of ordinary people. YouTube's autobiographical video diaries are a new form of self-presentation and an expression of a surrounding confessional culture. There is, in fact, a genre of videos with titles such as *I Confess, My Confession,* or *Confessions.* University student Rebecca Roth relates how easy online self-disclosure can be: 'It's surprisingly not hard to put details about yourself out to complete strangers. I'm not talking about details like your credit-card numbers. But, you know, details like "Oh, I feel this way about my mother."'[24]

Whether in the form of text, audio, or video, people find confessing online empowering. In her *10 Confessions* video a young woman, Fiona Liddell, admits to having doubted her friends in the past, not liking her body shape, being in love, trying to smoke to make herself look cooler, being afraid of the dark, having an unreasonable grudge, and wanting to change her name.[25] Fiona received a number of responses and, as a result, 'listened to a number of deep, personal confessions from different people who were thankful that I invited them to do so.'[26]Another confessional YouTube video made by a teenage boy combines the song 'Let the Bodies Hit the Floor' with pictures of his sister and her friends. The video is introduced with the words, 'I hate my sister to death and her friends.' YouTubers also confess their love for others. Another subgenre, spiritual confession, depicts individuals acknowledging that Jesus is their saviour. On YouTube we find all manner of personal confession. Western culture's ancient history of confessional practices helps to explain why video diarists find it so easy to divulge detailed and intimate accounts of their lives to distant strangers.

In his exploration of subjectivity in documentary film Michael Renov argues that video 'has had a special confessional vocation owing to its potential for intimacy and near-instantaneous feedback.' Renov sees confessional discourse as particularly well suited to video and asks why individuals 'choose to eviscerate themselves so profoundly for the camera.'[27] The western individual's propensity for confession has been famously explored by Michel Foucault in *History of Sexuality.* Foucault argues that embedded deep within the ancient and modern social institutions of the west was a formidable 'machinery of confession.' Institutions such as the Church, the military, science, and medicine, along with much of the bureaucratic apparatus of the modern state fostered individuals who were incited to produce a 'discourse of truth' about themselves.[28] As a result, many people feel quite comfortable telling friends and strangers about their private lives. Scientific experts promote psychologies of therapeutic healing, while religious authorities promote the theological necessity and psychological value of confession. Both modes of confession, notes Renov, require 'submission to authority.'[29] We could also name the market as another institution that promotes self-scrutiny. Advertising promotes a habit of constant self-measurement and comparison with others. Television and cinema are the great collective mirror. The Internet also stands among the latest in a series of novel representational technologies that promote self-scrutiny. Theories of confessional practices usually explore confession within the context of institutions and thus find authority and power relations to be at work. It remains to be seen how online confessional practices succumb to or evade authority and power.

There are plenty of reasons for Foucault to conclude that we have 'become a singularly confessing society ... Western man has become a confessing animal.'[30] Television has added to this contemporary obsession with confession. Mimi White argues that confession is a privileged and prominent part of contemporary television and appears across genres. She suggests that self-identity and social recognition 'hinge on participation in the process of mediated confession.'[31] This relationship between identity, recognition, and mediated confession certainly helps to explain why YouTube has become a giant virtual confession booth. In our hands the video camera acts as 'a kind of psychoanalytical stimulant which lets people do things they wouldn't otherwise do,' suggests Jean Rouch.[32] Renov similarly concludes that 'the presence of the camera or recorder is sufficient to spur self-revelation. In the case of video confessions, the virtual presence of a partner – the imagined other effectuated by the technology – turns out to be a more powerful facilitator of emotion than flesh-and-blood interlocutors.'[33] The YouTube audience acts as a virtual partner, an imagined friend, which generates a powerful impulse to confess.

Video diaries belong to the genre of confessional video and they also share some of the characteristics of ethnographic films, documentaries, personal art films, and reality television. The distinctions between all genres of film, television, and video practice have become increasingly blurred with each passing decade. This may be due to a general tendency towards convergence within media technologies, which in turn is fostering convergence within media practices. Boundaries that once separated practices are more porous, and thus categories of media practice and genre are constantly being redefined.

Across all mediums and genres there has been a general drift towards a more direct representation of intimate everyday life. It is as if all forms of media practice are converging on the self and the everyday. Media are converging on the self, the home is the centre of this convergence, and the video diary represents the nexus of media, self, and home. The most common location for the production of video diaries is the home.

In 1991 BBC's *Video Diaries* allowed amateurs to record video diaries of their lives and became the first example of a successful do-it-yourself documentary program to appeal to a mass audience. The series marked a significant innovation in television production, but some critics dismissed the show as a form of unhealthy self-absorption. Although careless critics might dismiss the amateur video diary and video bloggers as quasi-autobiographical, there is little reason for discounting the confessional and autobiographical nature of the work of many video diarists.

Video diaries are also known as videologs, video blogs, vlogs, vblogs, and vogs. Given the novelty of YouTube, very little research has been published on the use of domestic video for making online video diaries.[34] Video diaries predate YouTube by over thirty years, so comparisons between former and current video practices in this area will help to shed light on the significance of lonely girls and horny boys video blogging on the 'Tube. For example, different types of media affect the diary process.[35] How YouTube changes the diary process is as yet uncertain, but we do know that YouTube diary creation is a highly interactive process, owing to its public context and the audience feedback. Thus, online video diaries are distinguished by their high degree of reflexivity.

Researchers have long made use of participant-generated video diaries as a way of providing insight into an individual's experiences. Participant-generated videos are used by researchers in a wide variety of fields, because they offer 'a more direct understanding of people.'[36] These videos do not represent a pure version of the participant's experiences, however, because the research context influences the participant's performance. In one way or another researchers are involved in the making of a participant's video diary. Barbara Ellen Gibson describes this process as a form of co-production involving the researcher and the researched: 'Participants present themselves according to what is expected.'[37]

Videos produced in an explicit research context are affected by the participants' relationship with the project's researchers. A similar dynamic exists between online diarists and their audience. Feedback from the online audience, awareness of their expectations, fears of being wrongly judged – all this and more can be on the diarist's mind when he or she is in front of the camera. It would be a mistake to assume that YouTube's video diarists operate in a social void. We need to heed Pierre Bourdieu, who spoke of invisible structures that organize our actions.[38] Even in the privacy and solitude of their bedrooms, video diarists' performances are influenced by their awareness of other YouTube diaries, confessional television shows, and the feedback that they receive in the commentary section of their YouTube channels.

Self-made video diaries are not necessarily more authentic because they are less mediated, produced, or more spontaneous. Maria Pini argues that the 'physical absence of an observer makes little difference' to the video diarist because the diarist will internalize the imagined gaze of the other.[39] This other may be a researcher or, in the case of YouTube, may be the audience itself. When diarist Shel speaks of how it is not easy for her to open up and be fully herself because of the 'jackasses' on YouTube, we see one effect of the imagined gaze of the other.[40] We can thus speak of the surveillant and normative gaze of the online audience as always being co-present with the video diarist.

The Reflexive Online Self

All areas of media practice such as painting, photography, film, television, video, and writing have been analysed through the lens of a concept known as reflexivity. There is a wide range of definitions and theories behind the term, but here it will suffice to surf over a few of its uses. Reflexivity describes a state of higher self-awareness or mutual awareness that involves two or more people. Jay Ruby describes the reflexive condition that can exist between a media producer and an audience as awareness of the backstage processes of production.[41] This is also the general condition of the contemporary audience. We know quite a bit about the production techniques of film, television, and advertising.[42] Many YouTubers also know something about the process of using a video camera or producing a video clip.

Reflexivity expresses the tendency within YouTube to make one's self the subject matter, but it encompasses more than the self as subject. Whereas self-consciousness is a matter of 'I know' and communication is a matter of 'I tell you what I know,' reflexive communication incorporates an element of 'I know that you know that I know.' Ruby suggests that reflexivity undermines the attempt to pass off representations (media products) as 'authentic, truthful, objective records.' Here we arrive at a contradictory aspect of video diaries. On the one hand, the YouTube audience tends to see amateur video diaries as more real than what they find on television. On the other hand, the audience has greater awareness of the constructed nature of media artefacts and a diminishing belief in the 'objectively true reality of things.'[43] As the Internet intensifies the processes of reflexivity, could it be that the online audience is gaining a heightened state of reflexive awareness?

The reflexive condition of digital media is changing the relationship between private and public life. The realm of the public is increasingly invading the private.[44] We are moving from a social world where we constantly monitor ourselves to one where we constantly monitor each other's self-monitoring media practices. YouTubers are their own paparazzi.

The YouTube video diarist Ruthlyn provides us with an example of the reflexive character of online amateur diaries. Ruthlyn is a twenty-one-year-old black woman who lives in the southern United States and is married to a white U.S. Army enlisted soldier. Her YouTube video diary has over thirty clips that address issues arising as a result of her interracial marriage. In a seven-minute video Ruthlyn counters the misunderstandings of another YouTube video diarist who has made dozens of videos that discuss Ruthlyn.[45] She also maintains a collection of video responses to her own video diary and provides written and video commentary on the responses she receives and video diarists she has watched. This pattern of mutual monitoring and response is highly characteristic of YouTube's video diaries. Video responses on YouTube can be as short as grownup's twenty-eight-second *You Suck (video response)* wherein he simply says in a slurred voice, 'Congratulations, this [video] sucks and so do you. You suck and your video sucks. Congratulations to you.' YouTube's high degree of interactive communication is yet one more example of the Internet moving the audience from a strictly consumer-producer media paradigm to a form of reflexive expressiveness, although clearly not all expressions are equally thoughtful.[46]

Jay Smooth, producer of the YouTube channel illdoc1, made a tongue-in-cheek video called *Vlogging is Stupid.* Smooth talks about how he loves video blogging but also finds it very difficult: 'when I am not in front of a microphone I'm very quiet and introverted and for me to maintain this connection with you all requires me to project this performance version of myself which is a part of who I am but it is not who I usually am … the more response I get and the more connected I feel in the conversation with you all and the more pressure I put on myself to put on a show that is worthy of that connection.'[47] Smooth's reflexive monologue captures the complexity of subjectivity within online diaries. Here we see that there is not one self against which the truth or fiction of a performance is measured.

The self is framed and determined by its context and we express different selves within different contexts. The online diary is one of the most recent innovations in contexts for the performance of selfhood. It reflects the plural character of the self and provides insight into how we can generate plural versions of our experience. Video helps us to represent subjectivity as plural, intertextual, and interrelational. Multiple manifestations of the self are revealed in online relationships.

The online diary phenomenon challenges overstated claims about the death of the subject. In online diaries both the diarist and the audience encounter expressions of multiple subpersonalities. Each time we create a new context for self-expression, we also discover more aspects of selfhood. Instead of measuring online diaries against the notion of an authentic unitary self, we may be better off thinking in terms of authentic pluralism.[48] YouTubers are often seen struggling with the realization that they are not quite the same person seen in their videos. They also are seen creatively changing their mode of self-presentation through their video diaries and blogs.[49] The network context of online diaries fosters plurality and interactivity within confessional discourse.

Time and again through YouTube we encounter diarists who are interacting with their audiences. Susanna Egan proposes that autobiography be approached as an interactive genre: 'theorists have not thought about autobiography as an interactive genre even at the very simple level of what one might call "interpersonal relations."' Lauren, Shel, Fiona, Ruthlyn, and Smooth all are deeply involved with their audiences. Egan suggests that theorists remain 'several steps behind the practitioners of autobiography who have been experimenting for decades with plural versions of experience.'[50]

The case of an eighty-two-year-old video diarist, Peter, and his YouTube channel Geriatric1927 demonstrates how online confessional practices are a reflexive, interactive, community-based form of self-expression. Peter was one of YouTube's earliest elderly video bloggers. His first video, *first try,* was posted on 5 August 2006 and has been viewed over 2.8 million times. Peter describes himself as a widower who lives alone in the middle of England. He has posted over 188 video diary entries. Initially, Peter had a regular YouTube audience of between 50,000 and 200,000.

Throughout 2009 he continued to attract a few thousand viewers and had over 49,000 subscribers.

In Peter's 2008 Christmas video message he conveys his feelings to YouTubers: 'you are my extended family. I've watched so many of you over the years and you have become so familiar to me that I feel that you are also part of my family.' Peter explains that his experience of being a YouTube video diarist has opened up many opportunities that never would have been possible without his video blog. He makes new friends, is interviewed by the press, appears on BBC television, is invited to media conferences, and has met both the Queen of Jordan and the Queen of England. Over the years Peter has won acceptance within the YouTube community and is seen as a real individual.[51] The community was attracted to his efforts to tell his own story authentically.

An analysis of Peter's first eight videos by Dave Harley and Geraldine Fitzpatrick reveals that Peter's diary entries are a form of co-creation. Peter engaged in a dialogue with other YouTubers and, as a result, his spoken narrative 'emerges through an ongoing process of collaboration and co-creation with his viewers.'[52] In one video we see Peter directly address another YouTuber, Paul. Peter responds to Paul's request for more information about his life, his opinion of rude YouTubers (haters), and censorship.[53] This type of exchange is extremely common within YouTube. Two years later Peter relates that he has received many of his new YouTube friends in his own home.[54] Over time Peter learns from the YouTube community and becomes less shy and more technically skilled, and eventually he produces skits that represent situations that arise from YouTube. The stories video diarists tell and the way in which they tell them are influenced by and co-created with their audience. YouTube diarists are involved in an intensely social form of confessional narrative that is aptly described as a co-production.

The Facticity of Everyday Life

As an uncontrolled space of expression, the Internet has provided a natural home for our need to tell our own stories. Many observers of the social world have noted our collective fascination with the private and the personal. Because of this focus on the micro level of the everyday, the idiosyncratic, and the banal, we risk losing sight of macro level forces, such as class and ideology, that provide the context for video diaries. Jan Campbell and Janet Harbord argue that there is no need to suggest that 'a focus on micro narratives signifies an automatic evacuation of theory.'[55] Various strands of Marxist thought have buried ordinary experience under the weight of the macro to the point where the category of experience is eclipsed. Some scholars insist that the social system completely structures, determines, and reproduces the world and thus renders the category of actual experience irrelevant. If we conceptualize structures such as language or media as being all powerful, then the individual is seen as little more than a product of totalizing forces. Thus, postmodernists such as Jean Baudrillard also tend to see the real as an invalid category. Yet feminist theorists such as Elspeth Probyn counter that the self and 'the primacy of the real' are valid coordinates for discussing individual knowledge and experience.[56]

Campbell and Harbord make the intriguing suggestion that the inclusion of autobiographical material as a research subject forces us to confront the 'fictionality of theory with the facticity of everyday life.'[57] The study of the self and, by extension, the study of the other and reality itself are plagued by epistemological doubt. Contemporary scholarship betrays an ironic certainty regarding its own scepticism about self-knowledge.[58] In Mary Evans's *Missing Persons: The Impossibility of Auto/biography*, a title that speaks of the crisis of knowledge in academia, the claim is made that 'the prospect of "real" people ever emerging from the pages of auto/biography is very limited.' Evans's scepticism towards confessional discourse is such that she suggests that we should reclassify autobiography: 'its place on the library shelves is not with non-fiction but very much closer to fiction.'[59]

Evans's reasons for disbelieving that 'real' people can emerge from the autobiographical process provide an example of how the fictionality of theory can overcomplicate the facticity of everyday life. She notes, quite rightly, that conventional wisdom sees autobiography as 'becoming more true to the person,' yet for Evans this is merely another aspect of modernity's myth of progress. Evans makes a very valid point when she notes that the construction of the ideal self of the late twentieth century is overly individualistic and overlooks the role of the collective. Evans believes that autobiography is more fiction than fact because it reflects dominant modes of self-construction and self-representation and thus mirrors not reality or truth but 'prevailing moral discourses and perceptions of the acceptable extent of disclosure.'[60] Here autobiography and diary are seen as inauthentic because the speaker is constructed by the surrounding social order.

One cannot dismiss the influence of dominant modes and morals in the way YouTubers represent their personal realities. Yet only if such forces are thought to be totalizing and inescapable can we rule out the possibility of facticity within amateur online diaries. For every prevailing style of moral discourse and ethic of personal disclosure we will find numerous examples of the disruption of such conventions among amateur online diaries. Marxist-informed theories, along with liberal and conservative theorists, tend to displace the authentic into the realm of the past or the future. Yet is it not possible to be authentic in one's stance against modernity, oppression, or daily drudgery? Is there absolutely no authenticity or transparency to YouTubers' desire to be real within their video diaries?

YouTube's celebrity performance artist Chris Crocker provides an example of how fact and fiction can appear within the same moment of self-representation. Crocker's diary clip *Chris Crocker – Individuality/ Gender* (1,029,861 views) is an intensely individualistic and perhaps highly constructed performance that addresses the constructed nature of the self. Reflecting on the varieties of gendered performance, the gay, cross-dressing Crocker tells his audience, 'there is no such thing as *just* boy or *just* girl ... People ask why do you wear make up? I say I wear make up for the same reason I take a shit. My body tells me to take a shit, my mind tells me to put on makeup.' Is his performance mere fiction, a conventional representation of the unconventionally gay, an act of pure capitalistic individualism, a reflexive collective process of self-construction, a representation of prevailing moral discourse and acceptable disclosure, or an example of a repressed self seeking fuller self-expression in the unconstrained media environment of the Internet? Theory is constantly at risk of slipping into the domain of fictionality by our too quickly dismissing the possibility of the transparent, real, and authentic in even YouTube's more premeditated performance videos.

Here I will not wade any further into the deep waters of the debate on the ontological nature of mediated experience. Michael Renov once disavowed any interest in the 'ontological purity' of his claims for video confessions, and I will similarly sidestep the issue.[61] For now it is enough to note that YouTubers feel that online amateur video brings them closer to each other's experiences and presents reality differently from television. We can either endlessly complicate such claims or accept that people may actually have some insight into the nature of their own experiences.

Transformative Confessional Practice

Psychologists Wendy J. Wiener and George C. Rosenwald note that little attention has been paid to the psychology of keeping a diary.[62] This has not stopped many from suggesting that video diaries reflect excessive self-absorption. As Margaret Griffith observes, the primary focus of video diarists on the self 'gives rise to a narcissistic tone.'[63] Teenagers, in particular, tend to be excessively self-absorbed and this is reflected in their video diaries. Early psychology studies of diarists tended to emphasize their narcissistic aspect, probably as a result of the simple fact that, generally speaking, the diarist was 'talking' to only him- or herself.[64]

Wiener and Rosenwald identify a number of relational and developmental effects that follow from the process of diary keeping, or life writing, as it is sometimes called. These effects include helping adolescents to discover themselves as separate from their parents (facilitating differentiation), controlling and containing emotional experience, moulding overwhelming feelings into a presentable shape, dissolving boundaries between separate parts of the self, relinquishing emotional control and allowing repressed material to surface, charting progress towards self-improvement, and using the diary as a reflexive mirror to objectify and observe the self. Diaries also contain a 'partial representation of that which was in need of mastery as well as the record of its overcoming.'[65] As can be seen in this brief and incomplete survey, the psychological effects of diary-keeping can be substantial. The psychological effects that may be manifested also depend upon the individual's developmental stage. It remains to be seen how these psychological processes translate from the private context of written diaries to the public context of online diaries.

Returning to Shel's NannyCam diary clip, *the real me*, we witness the diarist struggling with the depth of her emotions and the fear of being misunderstood. She says, 'there is so much inside me that I could pour out but I do not know when it is appropriate.' We get a sense of her struggling to mould overwhelming feelings into a presentable shape. She discusses the ways she is similar to and different from her parents. She admits, 'maybe there is stuff that I do not want to admit to myself … because when I open it up it makes it real.' From Shel we get a strong sense of process and reflexivity. She is not unaffected by her online diary and the feedback she receives from the audience.

In his diary, *important message for everyone*, Calin, a young YouTuber tells us, 'the computer changed me, im way different then i was in grade 6 (im in grade 7).' Do online diaries change the diarist? My own experience of self-disclosure over the Internet via email, Gopher, FTP, Usenet, webcams, online games, websites, blogs, instant messaging, Facebook, and Twitter inclines me to agree with Wiener and Rosenwald's conclusion that diaries 'transform the diarist, however briefly or superficially.'[66] They did not investigate online diaries, but their conclusion certainly applies to the virtual mode of life writing: diaries help people to construct a personal sense of identity and can be transformative.

Online diarist Fiona Liddell described the YouTube effect in the following terms: 'Vlogging has certainly changed me in a number of different ways. When I first started my video diary, I had little confidence and was very unhappy with who I was. As time went on, I got several messages from people all around the globe telling me that I was an inspiration to them! YouTube has taken me through such an important journey in my youth. I have developed a confidence that I never knew I had. I've met some incredible people, and what's exciting is that I know that there are tons more that I still need to meet! I'm thankful to the YouTube community for their acceptance and for restoring my faith in myself.'[67] Viviane Serfaty describes online written diaries as 'thoroughly familiar and intensely new.'[68] The same thing can be said of video diaries: they are both familiar and new social experiences that are consequential.

YouTubers feel that their online diaries do change them. 'Remember, when you tell a girl she's ugly it takes ten more people telling her she's beautiful for that comment not to hurt anymore. No matter what we do, no matter how much we stand up, the haters are not going to go away,' relates one female diarist.[69] The self-esteem of online diarists gets tangled up in the Web of YouTube. Their diaries act as mirrors, but these are mirrors upon which other people can write nasty words. Egan's description of autobiographical film as a 'transformative intensification of lived experience' aptly applies to video diaries.[70] Likewise, Foucault saw confessional practices as producing 'intrinsic modifications' in individuals.[71]

According to Foucault, confession is a process that allows individuals to constitute themselves and determine their identities. We are what we confess. This helps to explain why we are so fascinated with personal revelation within social networks such as YouTube. YouTube's online diaries are seductive and appealing because they present confessions that the audience believes reveal the truth and 'our deepest selves.'[72] Confession is a normalizing process wherein individuals internalize the standards of acceptable behaviour. Given the extreme variety of confessional practices in the deinstitutionalized context of the Internet, it remains to be seen if Internet confessional practices will prove to be more normalizing than subversive.

Haters, neo-Nazis, jackasses, gangsters, gay-bashers, misogynists, and fools confess on YouTube and in so doing seek to create their own communities and encourage copycat thinking and behaviour. Confession often takes place within institutional settings and serves as a way of controlling deviance. In the absence of a controlling group or institutional setting, online confession may function more as a mode of normalizing deviance. Online diaries will change us, but not necessarily for the better.

There are rumours that the practice of autobiography is dying. Yet the Web provides ample demonstration of the staying power of the autobiographical genre.[73] Over twenty years ago Elizabeth W. Bruss claimed that autobiography was dying and there was 'no real cinematic equivalent for autobiography.'[74] Bruss saw the filmic mode as shattering the unity of the author and the text in autobiography, because film inserted 'the other' of the camera and the filmmaker between the author and the autobiographical self. Bruss was premature in declaring the imminent demise of autobiography. As Michael Renov notes, there is little reason for asserting that confessional discourse does not transfer well from written to video form.[75] Autobiography and diary have made an almost natural transition to the medium of video.

Summary

Video diaries are the punk version of television: they seek to be authentic and often defy conventional modes of self-representation. Diarists and their audiences value authenticity and transparency while also acknowledging the way the camera and the audience affect their self-presentation. The practice of diary and confession is moving from the domain of the private to the domain of the public and is now seen as a way to gain fame within media culture. The video diary culture of YouTube is intensifying the experience of reflexivity as we watch ourselves being watched by others. YouTubers experience their video practices as transformative and perhaps represent a new mode of self-construction, multiple selfhood, and identity maintenance.

Video diaries also represent a shift in popular tastes. We see a growing taste for the 'unmediated' and 'unedited' representation of the other (although mediation and editing remain). As Michael Z. Newman suggests, 'No longer is professionalism assumed to be the norm and standard of quality. The notion that do-it-yourself amateurism can stand on equal ground with media industry professionalism signals a democratic challenge to hierarchies of aesthetic value. And at the same time that amateur media is gaining ground, so is the communitarian alternative to traditional, top-down mass media distinctions between production and reception.'[76] Newman sees video blogging as a form of vernacular creative expression that relies on a network of cooperation. The real you within YouTube is fostering an emerging aesthetic value, an expression of a desire for something other than the highly produced, glossy reality of commercial media.

4 Women of the 'Tube

In the video *Is YouTube a Feminist Space?* twenty-year-old Morgan compiled clips from a variety of amateur videographers who discuss women and feminism.[1] This montage explored issues of misogyny, privacy, objectification, sex, bras, lesbianism, orgasms, and racism. One viewer responded by saying, 'You found some promising clips, but for each one of them I could show you three more disparaging, anti-feminist diatribes. We all assumed the web would be this "democratizing" space, but by the looks of things it's shaping up quite differently.' In another YouTube video, *This is What a Feminist Looks Like* (by Feminist), women are seen speaking positively about what feminism means to them. This montage was viewed over 125,000 times and garnered over 3,900 written comments. These videos represent a vast debate on feminism taking place on YouTube.

The future of women's voices on the Internet is in doubt. As recently as 2004 it was suggested that the Internet may follow the trajectory of earlier technologies and marginalize the voices of women.[2] However, this conclusion is hard to reconcile with the millions of women who are creating digital content and speaking through amateur online video. Women are opening up more spaces for self-representation. Earlier claims that women 'are rendered invisible, an absence, within dominant culture' do not automatically apply to the Internet, as this new media system lacks the institutional structures and male gatekeepers that defined the media systems of the twentieth century.[3] Nonetheless, we must take into account that YouTube's context is patriarchal capitalism and its misogynistic media culture. This context influences the ways women represent themselves and how the cyberspace audience respond to their voices and bodies. This chapter will survey women's use of online video to define themselves, develop relationships, and respond to misogyny within the 'Tube.

In the YouTube community there is an ongoing debate over women's place in society and on the 'Tube. For example, Jenna from Canada creates online videos as part of a dialogue on sexism.[4] Jenna relates: 'YouTube has been an incredible outlet for self-representation. I originally started making videos simply because I felt my feminist beliefs were being misrepresented by others who had been attempting to co-opt who I really was and what I meant to say for their own personal agendas. Putting myself on webcam and editing my own videos allowed me personal and creative control over my image that I otherwise didn't have.'[5] Thus, Jenna responds to the misrepresentation and co-option of her identity by creating her own online videos.

Another video blogger, Jessica Valenti, begins a clip by reflecting on how she has been harassed by misogynists since she went online and has been dealing with their 'sorry asses ever since then ... there certainly is a special breed of YouTube misogynist.'[6] One YouTuber responds to Jessica: 'Thank you for posting this! I'm so sick of these morons posting such disgusting hateful speak on anything feminist-related online. They needed to be called out.'[7] Women often resort to online video as a means of producing a counter-discourse. Typical of these exchanges, a debate also occurs within the video's written comments and video replies. Out of thirteen video replies by men to Jessica's speech eleven are hostile towards feminism. Jenna also must struggle with sexism: 'I must say that sexism (as well as other forms of oppression) here can be very rampant and the anonymous nature of YouTube doesn't help. There is a lot of harassment. I have had one online stalker in particular who just wouldn't leave me alone. It gets too much at times and people quit.'[8] YouTube has created a contested space where people engage in an intensely emotional battle over the proper place of women in society.

One of the earliest studies of the gender divide on YouTube found that men post video blogs more often than women (58 per cent to 33 per cent).[9] This study was conducted in 2006, so its conclusions reflect an early stage in the evolution of the YouTube audience. It also found no significant difference in the ages of male and female bloggers, in image and audio quality of their videos, or in the amount of editing by either sex. Significant gender differences within YouTube videos begin to appear when we consider viewer rates, interactivity, and subject matter. Women video bloggers 'were almost twice as likely to post vlogs [video blogs] that interacted with the YouTube community, and these types of vlogs received the most viewer "hits" ... Vlogs containing personal information were the second highest viewed videos.' The women of YouTube are more likely to blog about personal matters than men (60 per cent to 48 per cent).[10] These findings suggest that the YouTube audience is highly attracted to female video diarists who disclose personal information. Perhaps this is to be expected. Female YouTubers often draw large audiences, tend to focus on personal subject matter, and create highly interactive video diaries.

It is also noteworthy that women reported as frequently as men that they were part of the YouTube community.[11] This community space for self-representation is not marginalizing the voices of women. As a result, and Lonelygirl15 notwithstanding, female amateur videographers may also be seen to be authentic. Sidonie Smith and Julia Watson speak of how women's unpolished and unprocessed home movies emit 'an aura of authenticity.'[12] Yet many literary and film critics deny the authenticity of women's experiences and position them as imprisoned by the male gaze.[13] YouTube will provide further occasion for debating the authenticity of women's online self-expression.

Objectified Self-Representation

The use of YouTube by women invariably opens up their self-representations for erotic objectification by the male gaze.[14] One need only explore the countless amateur dance and bootie videos made by women to see sexism writ large in the written comments. Many of these bedroom dance videos are intended to invite the male gaze. Thus YouTube operates as a site where some women perform their objectification within a sexist media culture. Unfortunately, this category of objectified self-representation is densely populated by female 'tweens and teenagers who mimic the style of self-presentation found in misogynist and racist music videos.

Acts such as exotic dancing and stripping are not without contradictions. They are only two of many manifestations of how women perform their gender in the everyday world.[15] The temptation to dismiss bootie dancing, pole dancing, and striptease performances on YouTube as nothing more than objectification must be tempered by recent theories on these manifestations of neo-burlesque as an emerging form of feminist expression.[16] In bootie dancing and simulated striptease acts we witness women taking pleasure in their sexuality. Yet, as Catharine A.

MacKinnon points out, the issue of pleasure and how to get it 'rather than dominance and how to end it' is too often presented as the main issue that sexuality presents to feminism.[17]

As a form of resistance to male domination or simple self-expression, bootie dancing, pole dancing, and striptease performances on YouTube often do not move beyond the mainstream aesthetics of rap music and pornography. That YouTube returns over 190,000 hits on the search term 'bootie' does not indicate that here women have found a new means of empowerment and self-representation. Consider the home video *yay it's all GOOD* (13,869,330 views), which depicts the backside of a woman as she dances and removes her shorts to reveal a G-string. The 8,098 comments that accompany the video suggest little other than objectification by the male gaze. Does this video represent an act of empowerment by women or a 'failure to imagine their pleasures outside a dominant male economy'?[18]

YouTube and the Internet are changing how sexually explicit materials are produced, circulated, and consumed. Both the home (where most of these dance videos are made) and women's bodies are being redefined as sites of sexual pleasure. These new modes of amateur erotic performance represent acts of subjugation and moments of individual liberation through bodily pleasure.

Sexualized representations of women constitute one of YouTube's most densely populated visual themes. The pornography industry quickly adopted YouTube as a marketing vehicle. Videos marketing pornographic products receive millions of views on YouTube. While herein I do not explore the use of YouTube by professional strippers, exotic dancers, or the pornography industry (as my focus is on the amateur), there remains a very large field of amateur video practice in which women produce sexualized forms of self-expression. The availability of the Internet as a distribution medium and video as a low-cost/no-cost form of reproduction has led to a massive surge in the creation of amateur erotica. YouTube is home to all forms of sexual fetishes and marginal or underground sexual practices. Teenagers and young adults are often seen mimicking such behaviour in their home videos. There is even an entirely new genre of pseudo-pornography, produced mostly by male teenagers, that entices YouTube users to click on its videos only to find that the advertised sex act is not present. In another subgenre of amateur erotica, young women film themselves licking popsicles, bananas, and lollipops with obvious references to oral sex.

YouTube is quickly becoming a forum for distance education in both everyday and non-normative transgressive sexual practices. For example, hundreds of videos on YouTube feature women explaining how to give oral sex. Some of these videos, such as *Oral*, generate as many as 63 million views. More explicit material is quickly deleted by YouTube. Any rejected hardcore amateur pornography migrates to sites such as youporn. com.[19] YouTube acts as the equivalent of *Playboy* magazine – a site of soft-core pornography that can be consumed in the living rooms and bedrooms of America.

Whereas media and gender theory normally explores how the visual consumption of erotica and pornography affects the audience, we now find the audience as a major producer of sexualized performances. YouTube may accelerate the ongoing 'pornographication' of mainstream society. It also may contribute to the ongoing erosion of norms and regulations that, as Feona Attwood observes, are 'designed to keep the obscene at bay.'[20] Challenging as much of this material may be, the domestic production of amateur erotica and pornography by YouTubers is not a matter that can be readily dismissed as illegitimate, juvenile, immoral, and misogynist (although without question these judgments may often be correct). Pornography is becoming domesticated. It is being produced and consumed by women who have 'gained control of the means of production of erotic texts,' and this, as Jane Juffer suggests, is radicalizing sexual practice.[21] Amateur pornography on YouTube often is merely the public expression of everyday private sexual behaviour.

Five Awesome Girls

In January 2008 five American teenagers launched a YouTube channel called fiveawesomegirls. These girls decided that for an entire year each would make one video a week in order to get to know each other better. By the end of 2008 Kristina, Lauren, Kayley, Hayley, and Liane had made over 260 videos and provided an extraordinary demonstration of how women use YouTube to create a space for interaction, self-exploration, and self-expression. Their project has over 24,000 subscribers and their videos attract thousands of viewers.

In her first video Lauren explains that she 'looks like crap right now … You are probably going to realize that after a year of us doing this that sometimes I'm really lazy and I do not really like getting up and looking pretty.'[22] Lauren tells the audience various things about herself (she is a Harry Potter fan), invites feedback from the YouTube audience, and explains that as yet she does not know everyone in the fiveawesomegirls group very well. At fifteen, Kayley is the youngest member of the group. In her first video she speaks of her hair and the things she loves, such as Harry Potter and Disney (a year later Kayley will still express concern about her hair).[23] Liane's first video tells the audience that she is a musician and a Harry Potter fan. Hayley's first video explains that she has 'a very unhealthy obsession with Harry Potter' and is a vegetarian.[24] Kristina, also a Harry Potter fan, tells us that she likes reading, writing, shopping, and college.[25]

All the girls reveal considerable information about themselves and invite feedback from the audience. Over the course of the year they develop technical editing skills and produce increasingly sophisticated videos. They become far more confident in front of the camera. By the end of the year they have developed deep friendships with each other. These young women have used YouTube to create and maintain friendships across time and distance. Here we see more confirmation of the thesis that online relationships can lead to deeper and richer offline relationships.[26] The women of YouTube contradict earlier theories that presented the Internet user as a basement-dwelling loner.

The Strange Diary of Tricia Walsh-Smith

When a woman airs her grievances against her wealthy husband in a video made in their Park Avenue apartment, we see the boundary-crossing potential of online videos. In 2006 actress, playwright, and former Playboy bunny Tricia Walsh-Smith used YouTube to complain about the alleged bad behaviour of her husband, Philip Smith, a Broadway executive. The story of Tricia's online diary was reported in many newspapers in the English-speaking world.

Part diary, documentary, personal art film, and reality television show, her four YouTube videos not only tell an amazing tale but also highlight some of the ways that YouTube is changing how we represent our private lives. The *New York Magazine* described Tricia's online performance in the following terms: 'Her unblinking zombie eyes seemed to mirror some madness in the soul. Tricia Walsh-Smith was either a Jackie Collins character who'd short-circuited under the strain of overdrawn credit cards and lawyerly manipulations, or she was an astute video marketer aiming to mill disgrace into some rhinestone-studded rebirth à la Paris Hilton. One couldn't help but wonder how enamored she'd become of the media circus unfolding around her.'[27] Tricia certainly leaves the audience with much to wonder about.

This forty-nine-year-old actress was married to a seventy-four-year-old millionaire. When their marriage went sour, Tricia made a video that explains that her husband is taking action to have her evicted from their apartment.[28] Tricia speaks of how she was being tricked out of her divorce settlement. The camera follows her as she calls her husband's secretary on the telephone and explains that she never had sex with her husband, but she found Viagra, condoms, and pornographic movies in the apartment. In a later interview on Fox television Tricia explains to Greta Van Susteren: 'It's all about money, totally all about money.'[29]

Her video eventually was used by her husband's lawyers as grounds for divorce (the first legal instance of its kind). The Manhattan judge Harold Beeler called her initial video 'a calculated and callous campaign to embarrass and humiliate her husband … She has attempted to turn the life of her husband into a soap opera by directing, writing, acting in and producing a melodrama.'[30] This is the first instance of a divorce being aired via home movies on YouTube. As they involve a professional actress and were made by a professional video company, the status of her videos as amateur productions is questionable, but they nonetheless represent what may become a new genre of home video – reality divorce.

Tricia made two other videos that further explain her case and a music video that puts the story of her marriage to music (*Tricia Walsh-Smith Is Going Bonkers*). By positioning herself as a warrior, activist, and feminist, Tricia ignited an online debate about her motives and the legitimacy of feminism. The written comments that accompany her YouTube videos reveal a gendered reception to her plight. Comments by men tend to be very negative, while comments from women tend to be more supportive. Curiously, the majority of video replies to Tricia are by men and are negative. If you are a wealthy woman going through divorce and you air your dirty laundry via online video, YouTubers are not likely to be sympathetic.

On Being a Black Woman Online

Some of the most extraordinary videos on YouTube are produced by Black women who are struggling to define themselves. In an extensive subgenre of videos, the politics of the Black woman's body is discussed. YouTube reveals a community of Black women who interrogate the politics of Blackness and the effects of Black musicians and actors on their identity. It is noteworthy that within African diasporic discourse women play a central role in defining Black subjectivity. Black American feminists have successfully created 'a far more complex, textured, and inclusive notion of Black identity.'[31] This complexity is readily found within YouTube.

The Black community is embroiled in an emotionally charged debate over the proper way to represent and express 'Blackness.' There is no consensus within either the community or YouTube about what constitutes the acceptable form of Black identity.[32] One twenty-one-year-old woman, Delilah, created a YouTube channel to redefine 'what it means to be a woman and no less to be a strong black woman.' Her video, *Black Women Dropped the Ball (Attitude, Weave and Disrespect)* (152,782 views), calls on Black women to stop trying to be like white women. This video belongs to a large body of discourse that addresses the politics of racial identity.[33] Delilah positions Blackness as a natural form that has been degraded by contact with European culture. A young Black woman replied to Delilah with the video *I Got Weave In My Hair So Must Wanna B White* and defended a more multicultural understanding of Black identity.[34]

Another Black woman, Robin, delivers a deeply personal and moving monologue on how she has struggled to be both Black and a woman, 'I am not a ho, I am not a bitch, I am not a nappy-headed anything. I am a woman. I work hard at being a woman.'[35] Her video speaks of a lifetime of struggle against sexism and commercial representations of Black women. Both Delilah and Robin encourage their viewers to disengage from Black commercial artists who position women as sex objects. Douglas Kellner notes that 'one must often leave the mainstream to seek out the more radical and distinctive black voices.'[36] When we encounter these voices on YouTube we find that they inscribe complex, contested, and contradictory positions towards the mainstream itself.

Indigenous Women and Identity

There is a genre of YouTube videos that celebrates Native American women. Misha created a slide show music video entitled *Native American Women*.[37] The slide show comprises photographs that depict Native American women in traditional clothing. Numerous amateur videos also celebrate the beauty of Native American women. Most amateur representations of Native American women fall into three groups; slide shows celebrating Native American women, videos of cultural spectacles such as women's dance competitions, and videos made by teenagers or adults that depict everyday life in homes or on reservations. This brief analysis will focus on the third category of video.

Native American amateur videographers often depict everyday life in family settings. Men and women are seen singing together in kitchen parties. Children capture playful moments at home and on reservations in videos such as *rez life* and *Rez Kids*. The social conditions of reservation life often come through all too clearly. Home videos and slide shows often feature scenes of drinking at home or underage drinking. Videos also depict girls dancing or fighting.

I found two notable absences in the use of YouTube by Native American women. Thus far, confessional and autobiographical video diaries are extremely rare forms among them. There is also a distinct lack of debate over Native American identity within their amateur videos. In light of the small number of Native American women participating in YouTube, it would be dangerous to draw any conclusions, but we do see a less directly personal and political use of amateur video within this group (particularly when compared with Black women). This narrow slice of Native life is probably due to the relatively small number of Native Americans using YouTube at this point, as identity within the Native community is as contested and politicized as it is within the Black community.[38]

Barbie Girls

Women's identity is deeply entwined with commercial representations of femininity. Mattel's Barbie doll acts like a magnet within a media culture that attracts admiration and hostility and is a symbolic focal point for the debate over femininity.[39] The following will briefly explore how the women of YouTube respond to Barbie's role as a pink plastic princess of femininity.

In 1997 the music group Aqua released the song *Barbie Girl*. The song's lyrics position Barbie as a girl who says sexually suggestive things such as 'you can brush my hair, undress me everywhere.' Naturally, this song caught the attention of Mattel's lawyers, who tried but failed to sue Aqua for trademark infringement. Ironically, in 2009 Mattel used Aqua's tune in a Barbie commercial. On YouTube there are tens of thousands of mashups involving Mattel's Barbie doll and Aqua's *Barbie Girl* song. Some are quite benign, such as an amateur video of three girls dancing to the song. This entirely unremarkable performance has been viewed over 8 million times.[40] There are many such examples of young girls dancing to the Barbie song. However, the symbolism of these performances can be quite amazing. The video *Barbie Girl dance* depicts ten young performers

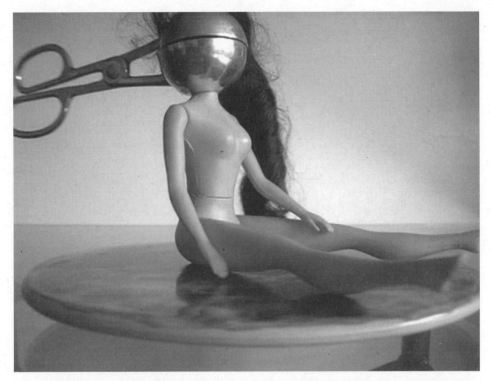

The YouTube video *Spinning Barbie*, by the author. Video artists use Barbie dolls to critique patriarchy and consumer culture.

dressed in pink dancing on a stage to *Barbie Girl*. Behind them is a giant American flag with the words 'Revolution Talent Competition' superimposed on it (the event is a dance competition). Revolution, capitalism, empire, and Barbie all share the stage at once. One suspects that Karl Marx would not be amused.

We do not know what is going on in the minds of young girls as they perform the Barbie song. Nonetheless, it is quite possible that their acts amount to the performance of hegemonic femininity. These amateur videos represent part of a lifelong engagement with commercial media culture wherein the female consumer identifies with certain notions of femininity and consumption. Or perhaps this is overinterpretation at a distance and the videos merely demonstrate that, as Cyndi Lauper once suggested, girls just want to have fun. In either case, there is something rather disturbing about watching children and teenagers enact sexual objectification through their performances on YouTube.

Another category of YouTube Barbie video challenges the meaning of Barbie and acts as a critique of normative conceptions of femininity. *Happy! Crazy! Fun! FEMINISM* uses Barbie and Ken dolls to explore domestic violence and the women's liberation movement.[41] Emily Hill's *barbie and the city (mature audiences only)* (1,194,100 views) created a clever appropriation that combined the voices of Samantha and Carrie from *Sex and the City* with scenes from the *Barbie Diaries* movie. The result shows Barbie reporting that her vagina is depressed and she has 'lost her orgasm.' Here Barbie is transformed into a representative of third-wave feminism.[42] Significantly, videos made by young men, such as *Slut Barbie* (1,550,041 views), *Barbie and Ken_Rough sex* (1,513,397 views), and *Barbie vs Psycho Ken* (924,956 views), tend to depict extreme violence and sexual degradation directed towards Barbie and female dolls.[43]

Through the use of Barbie on YouTube we see how women imitate dominant notions of femininity that originate in the marketplace, advertising, and mass media. We also see women challenge these notions and reject the stereotype offered by the pink princess of capitalism. Paulina's video *Not Barbie Dolls* uses a combination of text and images to reposition her identity as one of independence and strength. Yet there is a contradiction embedded in her identity claims; text in her video says, '"Girl" doesn't mean "pink barbie doll,"' yet she nonetheless draws on images of commercialized femininity from performers such as Avril Lavigne.[44] Here we see a young woman rejecting one concept of market-based femininity for another. The irony of her choice is seen in a 2008 commercial by Avril Lavigne for a Canon digital camera. Avril is seen with pink-tinted hair and pink fingernails, rummaging through a large pink trunk filled with pink consumer items, including a pink camera. Although she wears a black dress, the alternative rock star nonetheless remains a pink princess of consumption.

Fat Women

There is a wide array in women's styles of self-representation and in the issues they address (far more extensive than can be addressed herein). The following will briefly introduce three genres of women's online video practices that challenge mainstream conceptions of normalcy. The first of these genres is known as 'fat activism.'

It has been argued that women normally manage their weight to conform to dominant gender norms in the mass media.[45] Thus, it is quite significant to see the celebration of very large body types within YouTube. Joy Nash, a professional actress living in Los Angeles, made the video *A Fat Rant* in which she declares, 'I'm fat and that's ok.' Viewed over 1.4 million times, this video garnered more than 16,000 written comments and 74 video replies. While Nash is a professional actress, the replies to her video come from amateur videographers. These replies suggest that there is a community of YouTube members who resist the normative representation of femininity. Of course, some of the written and video replies are quite hateful.

The twenty-eight-year-old New Yorker, DeJadela, an amateur videographer, makes videos such as *OMG I Am Fat (Like I Didn't Already Know)* that also draw large audiences (508,810 views). DeJadela defends the necessity for all women to affirm a healthy self-love. Within YouTube communities we witness individuals affirming the identity and choices of other community members.

There is a forty-year history of fat activism and fat liberation. As a social movement, fat activism is 'increasingly self-confident and multi-faceted, and is beginning to form a widespread and viable community.'[46] Communities that are struggling for representation and self-confidence are turning to YouTube as a means of enacting a counter-discourse to prevailing prejudices.

Thin Women

Like the rest of the Internet, YouTube is home to a variety of extreme communities such as white-supremacy, pro-suicide, pro-drug use, self-mutilation, and pro-anorexia groups. Pro-anorexia community members use videos, also known as pro-anorexia and thinspiration, to argue their point of view and to proselytize. These videos attempt to promote the belief that eating disorders are valid choices. There are thousands of homemade pro-anorexia videos on YouTube, most of which are music videos that combine a song with images of very thin and unhealthy women.

One YouTube member, Sarah, posted a video entitled *Pro Ana* (150,970 views), which includes the text 'I support eating disorders.' Another video, *Thinspiration!* instructs viewers: 'Purge if you took more than 600 kcal. in half an hour, otherwise you'll only lose 50%.'[47] It remains unclear if these videos and similar pro-anorexia Web sites contribute to increased levels of eating disorders.[48] Many videos also provide a counter-discourse to the pro-anorexia point of view, such as *My Anorexic Story [updated]* (1,524,536 views).

The video practices of the pro-anorexia community are dominated by photomontages set to inspirational music, while voice-over, moving images, and direct-to-the-camera monologues are exceedingly rare. Virginia Heffernan suggests that these videos reflect the typical feelings of a young woman who is 'furtive, prolific, deeply melancholy, proud of her sacrifices, furious at her family's various offenses, frustrated with her body and protective of her supreme right to destroy herself.'[49]

It is insufficient to simply accept medical and psychiatric claims that pathologize pro-anorexia behaviour.[50] Such behaviour, while ultimately harmful, embodies complex attitudes towards society and mainstream representations of femininity. Anorexic behaviour can represent a rejection of patriarchal feminine ideals.[51] Much of the online response to pro-anorexia videos may only silence women. Debra Ferreday suggests that what is also at stake in the rejection of pro-anorexia narratives is the ejection of outsiders 'in order to reinstate the notion of consensus through the suppression of some forms of difference.'[52] The surrounding YouTube community tends to reject the values and discourse of extreme communities within its midst, yet there are complexities and contradictions that must be acknowledged.

Violent Women

Young girls are harshly beating each other, filming the action, and uploading the videos to YouTube. One of the more infamous cases involved six female teens beating sixteen-year-old Victoria Lindsay in March 2008. There are hundreds of videos on YouTube that discuss the event. Amateur videographers have made videos that show the faces and names of the six and debate their actions. Girl-on-girl fighting involves a complex and contradictory form of gendered performance 'that both perpetuates and challenges the usual notions of masculinity and femininity and the differential power associated with these discourses.'[53]

Curiously, there are many statements in the press and in academic literature that quote YouTube spokespersons or guidelines that claim: 'Real violence on YouTube is not allowed. If a video shows someone getting "hurt, attacked, or humiliated," it will be removed.'[54] This is public relations nonsense. Fairly typical of the fight genre, the video *Huge Black Girl Fight* has been on YouTube since March 2008 and has been viewed over 2 million times. The video depicts a street fight between a group of young Black teenagers. Videos depicting girls fighting are common enough to have their own tagging taxonomy on YouTube (e.g., chick fights 'gone bad,' 'in the hood,' 'after school,' 'in Hawaii,' 'in California'). A related video genre called 'hood fights' is also very common.

Police departments across western nations are reporting increases in violent crime among young people. According to Paul Mazerolle, director of Australia's Griffith University Centre for Ethics, Law, Justice and Governance, 'youth violence is still largely driven by young males, but young females are trending upwards ... [websites are generating] competition and encouraged them to look at ways of gaining status. Young people want to demonstrate superiority and toughness. That's why we've seen a proliferation of things like the videotaping of violent confrontations.'[55] The debate surrounding violence and the media remains unresolved, yet we are seeing something quite new: the amateur representation of street-level and domestic violence. Frequently, it is women both engaged in violence and recording violence. On YouTube the representation of girl-on-girl violence is often combined with the discourse of pornography. The names of the individuals involved in the fight are occasionally recorded within the videos or descriptive comments. This practice ensures that their identities will remain wounded long after the physical scars heal.

Productive, Active, and Strong Feminine Online Voices

The varied and numerous female voices within YouTube challenge earlier claims that the Internet is dominated by men. In 1995 Dale Spender confidently declared, 'Make no mistake about it, the Internet is male territory.'[56] Likewise, Margie Wylie proclaimed that electronic networks such as the Internet will soon be 'no place for women.'[57] Yet these assessments suffer from having been made far too early in the history of the Internet. In the late 1980s and early 1990s the Internet was certainly dominated by men, but this is becoming increasingly less the case.[58] The American Internet population, for example, has reached gender parity even among senior citizens.[59] China's Internet is fast approaching gender parity.[60] Within less developed countries the Internet gender ratio skews in favour of men, but this disparity also will change over time; males accounted for 60 per cent of the YouTube community in 2006 but now account for only 51 per cent.[61]

Two separate surveys concluded that men prefer amateur videos and women prefer professional videos.[62] The issue of gendered usage of the Internet cannot be settled by population and usage statistics, as we also must consider what the sexes are doing online, who is contributing what types of content, and the visibility of different types of content.[63] One female Internet user's comments on these studies say much about the character of YouTube as a gendered territory: 'although I use YouTube regularly too, I find the atmosphere – particularly the comments – on sites like YouTube quite juvenile and sexist, so it doesn't really make me want to stay on there for long.'[64]

Although Internet gender ratios vary considerably across income and ethnicity, women and girls are now among the most prolific users. In the United States and Britain girls outpace boys in their use of blogs and social networks. The reason for this would be obvious to many parents. Anthony White notes, 'If you look at young girls, they do more communicating than young boys and that's what they are doing on the web.'[65] Nonetheless, as much as the Internet once was a geeky male domain, it is now fast becoming feminized.

Women's voices within YouTube defy men's cultural domination over the Internet. Yet this situation must not be overstated. The women of YouTube are harassed, ridiculed, stalked, and subject to violent and vulgar speech. Video blogger Jenna from California reflects on how women on the street get 'heckled all the time' and relates that, because men harass her: 'most of the time I get on YouTube I do not even blog anymore.'[66] For Jenna, the street and YouTube provide the same experience of sexism.

Given all the contradictions that surround women's self-representation and self-expression on YouTube, Liesbet van Zoonen's assertion that Internet communications 'primarily tend to construct women as online consumers' oversimplifies the complexity of the matter.[67] The women of YouTube are engaged in the consumption and performance of market-based notions of femininity, but they also actively resist such dominant role models. The Internet also constructs women as online producers of meanings. Can we rightly say that the Internet *primarily* constructs women as consumers when women are also extremely active in building relationships and creating online cultural content?

The women of YouTube also act as a challenge to van Zoonen's assertion that the male domination of the Internet is increasing.[68] This claim overlooks the fact that the women of YouTube are teaching themselves to speak out against sexism. Jenna from Canada reflects on her experience of creating online videos: 'while there is negative feedback YouTube still has been great for situating where I am within society, for getting tougher skin and learning from past projects. I believe I am a better speaker and have more confidence because of my participation here.'[69] Jenna's experience of video-based online expression demonstrates how the women of YouTube deny male domination and control over online cultural production. Women like Jenna are developing video-based communication skills, gaining the confidence to speak out, encouraging each other to participate in dialogue, and building networks of support and friendship. One might go so far as to say that the feminization of the Internet is increasing.

It is time to put to rest the question asked by Tracy L.M. Kennedy and many others about whether the Internet 'gives "voice" to women and women's issues.'[70] It does. On the role of the Internet as a new type of space for women's voices Ann Travers notes that, while feminists have remained largely invisible in the public sphere, they gain more visibility in the counterpublics of cyberspace.[71] Through arenas such as YouTube women have not simply gained a new place to speak, but may actually be gaining greater visibility and broader audiences. In all areas of the globe the level of women's participation in the Internet is increasing.

Anthropologist Micaela Di Leonardo notes that the widely accepted thesis that women's voices are muted or subordinate must not be overstated. There are many different types of speech positions that women can assume. Women have a wide 'array of verbal strategies and genres' and can be 'considerably more articulate and more actively oppositional' than conventional models of gendered speech power allow. Di Leonardo makes the cogent point that domination and resistance 'are matters of interactional practice as well as structure.'[72] Online amateur video practices represent both a novel set of interactional practices and a novel structure of communication. We must be wary of assuming that domination in other mediums and social contexts will directly translate into domination in this new media context. One of the more certain ethnographic facts of YouTube is that women are speaking. The next most certain ethnographic fact is that some of the men of YouTube are trying to silence women's voices. This situation does not readily fit into the binary subordinate/dominant model.

There is a parallel between the way media theorists treat women as audiences of mass media and as Internet users. In *Watching Dallas* Ien Ang notes that both the representation of women in mass culture and women's reception of these images are viewed paternalistically. As a result, members of the female audience are reduced to passive, unwitting, and pathetic victims of the commercial culture industry.[73] New media theorists deploy the same type of paternalism when they reduce women to subordinate positions and claim that they are primarily being constructed by marketplace ideologies and patriarchal fantasies. Undeniably, such regressive forces are at work, but the progressive position of the women of YouTube as active producers of their own content and identity must also be recognized.

There is a distinct tendency to position corporations and men, the marketplace and masculinity, as dominant forces in the Internet, yet the gendered nature of participation and production within the Internet is rapidly changing and the evidence constantly shifting. There is no one dominant ideology or gender that is inscribed within the Internet. As Juliann Emmons Allison notes, both values that 'perpetuate unequal gender relations' and 'a substantial voice of opposition to those values as well' can be found within the Internet.[74] There are spaces in cyberspace where the male voice and corporations dominate. There are feminine online spaces and places that privilege the production of feminine texts.

Although some media theorists position the Internet as a masculine domain,[75] perhaps it is more like a soap opera. Ang describes soap operas as feminine texts that lack an ideological consensus and are filled with ambivalence and contradiction.[76] As an emergent domain the Internet lacks ideological consensus and is filled with ambivalence and contradiction. Within this media environment the women of YouTube are not merely given a modest degree of freedom to construct their own meanings. This is not an audience restricted to watching stories from a male-dominated, corporate storytelling machine. This is not an audience relegated to the weak position of interpreting someone else's stories. The women of YouTube are producers of their own stories. The ability to make our own stories is not the power of the weak; it is and always has been the power of the strong. YouTube may prove to be a natural environment for feminine texts and women who, as Valerie Walkerdine has observed, spend a lifetime performing.[77]

Women's bodies and sexuality are often cited as their main means to power, but this old theory needs revision. In a new media environment that privileges those who produce content women gain a new type of power. However, the new power of amateur online video and the interactive dialogue and relationship building that accompany it may not be realized or experienced in exactly the same way by both men and women. There are gendered differences between the way males and females communicate and the ways these differences are manifested are strongly dependent upon the social context.[78] Further research will be needed to determine how YouTube affects gendered styles of communication.

Some studies indicate that the similarities between communication and interaction styles of men and women outweigh the differences.[79] Therefore, we must be cautious about rushing to conclusions about how new communication environments change the balance of power between the sexes. YouTube has restructured the media environment and enhanced the productive opportunities available to both sexes. YouTube has also added a new mode of interaction between and among the sexes. The production of globally accessible amateur video defines this new mode of interaction, which includes video-based and text-supplemented modes of dialogue and relationship building at a distance. Such a change in the context of conversational interaction may also change how the sexes interact, given that interaction has always been conditioned by context.

Summary

In all these instances of women's representational practices there exist contradictory forces that reinforce and undermine dominant notions of femininity. Women's online video practices challenge some of the cherished assumptions within feminist media theory about the domination of the Internet by male voices. Video practices are gendered: women are producing more personal and interactive videos. Women's video practices can be highly social and have an aura of authenticity. In all instances they face resistance and objectification from some male community members.

In *Watching Dallas* Ien Ang carefully qualified her analysis by noting that no single media experience is unambiguous: 'it is always ambivalent and contradictory.'[80] We see this again in women's ambivalent and contradictory experience of YouTube. In her video *Re: Being a Chick on YouTube* Melissa Jenna Compagucci looks forward to a time when hate speech and sexist comments will diminish on YouTube and 'girls will not be so afraid to get involved in YouTube and other people will respect YouTube more as a forum to exchange ideas.'[81] We can hope that Melissa and the women of YouTube will not be disappointed.

In many cases it is clear that women derive pleasure from being amateur videographers. Whereas Marxists would position cultural production as being subjected to the economic principle of profit, on the Internet we find spaces of amateur cultural production that express not profit motives but the pleasure principle. Girls just want to have fun. In many instances they just want the freedom to express themselves in a safe environment, free from domination, sexism, and hate. In the commercial cultural industry pleasure was derived from the process of consumption. On YouTube we find pleasure results from the productive activity of amateur videographers. Women videographers both experience pleasure and give pleasure as producers of their own stories. Private and public pleasures collide and result in community and contested identities. Ang suggests that 'popular pleasure is first and foremost a pleasure of recognition.'[82] On YouTube women recognize themselves through and take pleasure in each other's video work.

5 The YouTube Community

The phrase 'Internet community' occurs over 10 million times on the Internet and is found in tens of thousands of academic articles. The idea of community has long been attached to the virtual realm but is not without controversy.[1] As is to be expected, there is no single, universally held definition of community and some scholars question the possibility of real community in virtual contexts. As Maria Bakardjieva notes, the concept of virtual community 'has led analysis into a not particularly productive ideological exchange disputing the possibility that genuine community can be sustained through computer networks.'[2] Sceptics aside, there remains the simple fact that many Internet users see themselves as part of a community; this is particularly true of YouTubers. Tens of thousands of amateur videos that address the YouTube community are uploaded every day.[3]

Perhaps not surprisingly, participation in online groups leads to a psychological sense of community.[4] People can be deeply engaged in online communities. Curiously, Internet communities have 'received mostly negative appraisals among psychologists.'[5] Nonetheless, members of virtual communities tend to believe that they know the personalities of other community members and that they both experience and observe more personal relationships than do members of offline (face-to-face) communities.[6]

Anthropology provides a more balanced assessment of online community than does psychology. Anthropology is a discipline that specializes in exploring communities. Samuel M. Wilson and Leighton C. Peterson suggest that anthropology is 'uniquely suited' for the study of the Internet. They note that for some time anthropology has found community to be 'a difficult focus of study.'[7] This difficulty certainly has been shared by new media theorists.

As has also been noted by postmodernists, the notion of community can often imply a false coherence and consensus. Whereas communities were once seen as bounded, isolated, and homogeneous, we now find them to be highly interconnected, heterogeneous, and rife with discord and dispute. This fluid and unstable character of contemporary communities leads Wilson and Peterson to acknowledge that 'individuals within any community are simultaneously part of other interacting communities, societies, or cultures.'[8] This observation reflects the conditions we find in Internet communities.

Wilson and Peterson remind us that there is little point in defining community as something that is based in face-to-face interactions. This line of argument was 'effectively challenged long ago.'[9] As Benedict Anderson made clear in his seminal work, *Imagined Communities*, 'Communities are to be distinguished, not by their falsity/genuineness, but by the style in which they are imagined.'[10] To ask if there is a real community behind virtual relations is to miss the point of what communities are – a shared imagining. By using the false starting point of envisioning community as based in face-to-face relations, we end up unnecessarily problematizing the lack of face-to-face interaction in cyberspace. Communities have always been partially constructed by a shared imagination, and the Internet is playing an increasing role in mediating the imagination. Amateur online video practices are also putting millions of faces online.

From the anthropological point of view, dichotomies of online/offline, real and virtual, and individual and collective tend to be overstated. Thus, Wilson and Peterson suggest, 'the distinction of real and imagined or virtual community is not a useful one.'[11] The key issue here is that online communities often exist in some fashion as a continuation of offline communities. Whereas it was once feared that the Internet would weaken community ties, there is a growing body of evidence that suggests that involvement in online communities can enhance existing local relationships.[12] Of course, other research suggests that the Internet weakens family ties, since there has been a sharp decline in the amount of time American families spend with other family members.[13] Nonetheless, there appears to be a strong correlation between offline and online sociability: 'Rather than weakening other forms of community, those who are more active offline are more active online – and vice versa.'[14] The offline and online worlds tend to be deeply interconnected.

One way of defining community is through the notion of shared interests. Fan communities, for example, share an interest in a particular story, television show, author, director, or celebrity (and so forth). On YouTube we find groups of individuals who interact around shared interests. Yet there is more to online communities than interest, as Steven G. Jones points out. Online communities not only are 'composed of people who are necessarily connected, even by interest, but are rather groupings of people headed in the same direction, for a time.'[15] Jan Fernback's argument that online communities of interest are 'closed places … that lack a social role in the larger collectivity' raises the spectre of the online/offline dichotomy against which Wilson and Peterson advise.[16] Yes, some online communities are disconnected, self-seeking, atomized, and have shallow roots, but the same can be said of some face-to-face communities.

Research on online communities often takes a small slice of online community and measures it against a narrow model of face-to-face communities and some general pronouncement is rendered by the theorist. Thus, we see Fernbank confidently proclaim that what is missing in virtual communities 'is the sense of individuality that can operate within the collectivity.' Only if it supports something called 'true democracy' and 'actualized individual identity' can the Internet manifest community.[17] Online community therefore comes to be measured against some parochial western notions of democracy and identity and is found wanting. It is indeed odd that so much intellectual activity has gone into discounting what many people claim – that the Internet is home to our communities.

Another feature suggesting that YouTube provides a space for multiple communities is that people identify with it. They invest time in it and develop relationships through it. A community, virtual or real, is something that people care about. As they care about their communities, they also are seen to defend them and debate their values and goals. Indeed, according to the anthropologist Mary Douglas, it is this very process of debate that constitutes a community and perpetuates its very existence.[18] To a certain extent, then, where there is disagreement and debate, there we find community. If we want to know what types of community YouTube has enabled, one of our best sources will be the internal debates and controversies that are found among YouTube's members.

YouTube members engage in debates on just about everything under the sun, but some issues act as significant indicators of various forms of community identity within YouTube. Like a neighbourhood community's dependence on local and national levels of government for its existence, amateur videographers depend on the goodwill of the YouTube corporation. This dependence leads to tension, as some members develop a false sense of entitlement to YouTube's services.[19] This sense of entitlement is aggravated by the corporation's acts of censorship and account cancellations. YouTube's own goal of generating revenue through advertising and sponsored content from professional performers such as Oprah Winfrey also generates resentment among some members. These areas of conflict fall into two broad categories: YouTube as a village cop – a regulator, patron, and landlord of this virtual village – and the perceived intrusions by commercial production studios, television shows, and celebrities. Along with YouTube as patron and celebrities as privileged members, the following will explore how fan communities use parody, how the community responds to bad behaviour (haters, flaggers, and spammers), and other patterns in YouTube use.

The Village Cop

In common with the search engine Google, YouTube is continually faced with individuals and companies that try to 'game the system' by taking advantage of YouTube's technical design to attract more visitors. This becomes an issue for the YouTube corporation because sexually suggestive images and words are frequently used to attract people to a video. Regardless of YouTube's own rhetoric about not allowing pornography, violence, and racism, there are hundreds of thousands (if not millions) of videos and written comments on YouTube that are exactly that – pornographic, violent, and racist. Some of these videos advertise pornographic websites and some are amateur videos that use pornographic titles, descriptions, or images in their thumbnails to entice the YouTube user to click on their video. The YouTube corporation claims that 'videos featuring pornographic images or sex acts are always removed from the site when they're flagged,' but this is often not the case.[20] While one may quibble over whether or not some of this material is sexually suggestive (and thus allowed) or outright pornographic, it is hard to see how video titles such as *Teenager Gets Fucked and Gives a Blow* or descriptions such as 'sex hard porn xxx blowjob vagina handjob pussy ass cumshot brazzers porno uncensored' fit with YouTube's content policies.[21]

As the village cop (and judge, jury, and executioner), the YouTube corporation occasionally changes how it regulates content. As Jean Burgess and Joshua Green observe, YouTube is under pressure to make the community 'more palatable to the public and the advertisers.'[22] In

December 2008 YouTube announced stricter rules governing sexually suggestive material. This material will be available only to viewers who are eighteen or older (although there is no way to stop a user from lying about his or her age!). Videos that do have sexually suggestive material in them will be demoted from pages such as 'Most Viewed' and 'Top Favorite.' The algorithm that automatically generates video thumbnails was changed to make it more difficult for people to use sexually suggestive thumbnails to attract viewers. The corporation also stated that it has 'always prohibited folks from attempting to game view counts by entering misleading information in video descriptions, tags, titles, and other metadata' and will insist on more accurate video information.[23] These various actions have done little to change the character of YouTube's content. Sexist, violent, and racist videos and comments are easily found on YouTube.

Using sexually suggestive video thumbnails, tags, titles, and descriptions to mislead the audience remains a very common practice. YouTube's assertion that it has always prohibited the use of misleading information in video metadata may be true, but it is equally true that misleading and pornographic metadata proliferate on YouTube. Consider the example of one of the most viewed YouTube videos, *XXX PORN XXX*, by Abolish-TheSenateOrg (viewed 111, 406,343 times). This video has a sexually suggestive thumbnail image. Its descriptive text is 'MILF gets Banged,' and its tags are equally pornographic.[24] The video itself is a forty-six-second slide show about abolishing the United States Senate. There are hundreds of videos with the same title and over 316,000 videos that have the word 'porn' in either their title or description.

Clearly, the rhetoric that the YouTube corporation uses to describe its regulatory practices is at odds with the actual character of YouTube. In common with many other Internet businesses that provide social networking platforms, there is a gap between the moral claims the corporation makes about its online property and its actual state. This gap between the corporation's public relations discourse and the YouTube community's actual practices may be due to the corporation's need to provide itself with a legal defence against possible lawsuits. It may also be due to its need to appear as an advertiser- and family-friendly environment.

Business media analysts often provide a conduit for the corporation's claims. Thus, we see Allen Weiner, a digital media analyst, assure business readers that YouTube's executives are 'beginning to make YouTube a far more clean, well-lit place that will attract advertisers.'[25] This optimism is exactly the type of message you would want to hear if you had invested in YouTube or were thinking of using it as an advertising platform. YouTube's words and actions should always be interpreted in light of its primary position as an advertising platform and its need to keep its main client – corporations – reasonably happy.

YouTube's new regulations generated concern among some members of the community. In response to the new regulations a thirty-four-year-old man made a video entitled *Youtube is Dead*. Within three weeks this video attracted 82,000 views, 61 video replies, and 1,541 written comments. This YouTube member felt that the new regulations amounted to censorship: 'I think YouTube wants more civilized and homogenized videos so that they can increase their advertising revenue.' Some YouTube members made videos encouraging others to write to the corporation and protest the changes. Others encouraged YouTubers to boycott the site for a day. These protests and internal discussions have the character of a tempest in a teapot. They represent real fears and suggest possible future outcomes of the corporation's actions, but very little by way of substantial changes in the content and culture of YouTube can be detected at this stage.

One YouTuber responded to the corporation's new algorithm for demoting improper videos by posting a video entitled *Algorithmically Demote This!* The video consisted solely of the names of other Internet video hosting sites such as Metcalfe, LiveVideo, and LiveLeak.[26] The meaning was clear: YouTube has competition and, if the corporation fails to host a video, it will find a place elsewhere on the Internet (which is more or less the case). In November 2009 there were over 48,100 videos that mentioned YouTube and censorship and 2,980 videos containing the phrase 'YouTube sucks' in their titles or descriptions.

A video rejected by YouTube can be slightly altered and once again uploaded to YouTube or instead uploaded to countless other video, Web, blog, or social network sites. It remains to be seen if the corporation's policies will have any significant effect on online content and amateur video practices. As others have noted, amateur videographers 'are not captive to YouTube's architecture.'[27] Not only are they not captives of the 'Tube, they are also very adept at working around its controlling mechanisms.

· The fear of censorship within the Internet community has always been greater than its actual effectiveness. It is noteworthy that YouTube tolerates videos that discuss boycotting YouTube and that accuse it of censorship. An individual who has had an account suspended or deleted can acquire a new account and use it to complain about having the old account removed. We also see people uploading other individuals' banned videos to ensure that they are still on YouTube. In some cases YouTube has reinstalled a banned video in response to complaints of unfairness from YouTube members. As the village cop, the YouTube corporation gets caught up in complex webs of accusations and counter-accusations by YouTube members who are trying to ban each other. These types of dispute often centre on political and religious topics. It may be the case that more acts of censorship are initiated by YouTube members than by the corporation itself.

All in all, the YouTube corporation has thus far demonstrated a remarkable degree of tolerance. Burgess and Green describe YouTube as a patron that provides both 'supporting and constraining mechanisms.'[28] It remains to be seen if YouTube will continue to play the role of a village cop or transform into an authoritarian tyrant.

A Community of Pirates?

YouTube is faced with numerous lawsuits, including a $1 billion claim by Viacom that alleges that the corporation is facilitating copyright infringement.[29] It has responded to some of these lawsuits by deleting hundreds of thousands of videos. These occasional mass deletions have had little effect on the type of content available on YouTube. What the *New York Times* told its readers in 2007 remains true today: 'No one knows exactly how much Hollywood-derived content is uploaded to the site without the studios'·consent, but academics and media executives estimate it could be anywhere from 30 percent to 70 percent.'[30] The exact percentage depends upon how 'fair use' is measured. Since media corporations tend to interpret fair use as copyright violations, from their warped point of view YouTube is a community of pirates. Whatever the actual amount, it remains true that corporations have lost control over their digital property and may not be able to fully re-establish control within the Internet.

On the Internet individuals do not have total freedom and corporations and governments do not have total control. YouTube is private real estate controlled by one corporation. Thus, there is a much higher level of control within YouTube than is found within the surrounding domain of the Internet. An MIT Free Culture project called YouTomb (youtomb. mit.edu) has engaged in an interesting experiment, which tracks videos removed from YouTube for alleged copyright violations. Their intention

is to shed light on YouTube's less than transparent process for responding to Digital Millennium Copyright Act (DMCA) takedown notifications
and provide a resource for YouTubers who have had their videos wrong-
fully removed.[31] YouTomb provides an example of how the Internet can
act as an oppositional force.

Individuals and groups use the Internet to organize and facilitate
resistance to the claims and actions of corporations and governments.
As much as digital technology allows corporations to identify copyright
violations, the Internet's enabling characteristics allow individuals to
monitor and challenge violations of fair use. YouTube also enables the
mobilization of individuals in the ongoing information war between
governments and citizens, corporations and consumers.[32] In July 2009
there were approximately 4,660 YouTube videos that referred to the
DMCA. Amateur videographers have made videos explaining how to file
a DMCA notice and how to contest a DMCA takedown action.[33] Other
amateur-produced videos call for DMCA reform and criticize YouTube's
DMCA processes.

When corporations tread upon individuals' fair use rights they stimulate collective political action directed towards reforming DMCA
practices and copyright law. The YouTube community and amateur videographers may also foster changes in copyright law because 'fair use doctrine evolves with creative practice.'[34] As YouTubers creatively expand
the uses of copyrighted works, they will likewise encourage changes in
legal theory and practice. Given that copyrights are largely irrelevant in
the mind of the average amateur videographer and that most copyright
infringements go unchallenged within the Internet, copyright law may
increasingly become unenforceable. It is not completely enforceable
'because copyright is not really a concern in the public mind.'[35] As K.
Matthew Dames notes, 'Neither large, corporate copyright owners nor
Congress can expect Joe Citizen to abide by the interpretation of today's
copyright law because that interpretation is unreasonable to the point
of being stupid.'[36] Mass involvement in amateur video production may
provide the citizenry with an education in the inequities and inanities of
the law.

Will the YouTube corporation itself act as champion for fairer inter-
pretations of the DMCA and copyright law? As Jeffrey C. Brown notes,
YouTube 'is continuing to pursue efforts to minimize its exposure to
copyright infringement liability, while increasing its potential advertising
revenue.' These priorities and YouTube's partnerships with major entertainment corporations, movie studios, and television networks over-

ride concerns for individual fair use and suggest 'a long-term business strategy that depends less on user submissions and more on synergistic collaboration with the recording and entertainment industries.'[37] Yet YouTube will always need the willing participation of amateur videographers if it is to remain competitive.

From the earliest days of the World Wide Web there have been concerns that freedom of expression and creativity in the Internet would be shut down as a result of corporations extending control over their intellectual property. Tarleton Gillespie argues that the Internet imposes 'tight controls' on communication and digital culture, but this claim cannot explain the everyday uses of the Internet by ordinary people.[38] Communication, digital piracy, copyright, and intellectual property rights within the Internet cannot accurately be described as subject to tight controls. From its earliest days the Internet has consistently expanded communicative freedom and cultural production. DMCA takedown actions on YouTube notwithstanding, online expressive freedom and digital culture are not being wired shut.

Regardless of the outcome of various lawsuits against the corporation, much expressive activity within YouTube does not tread upon corporate intellectual property. As Kurt Hunt notes, 'there is no question that a great deal of YouTube content is non-infringing.' Hunt's legal analysis concludes that 'many YouTube users should be considered as having a strong fair use defense.' Hunt also recognizes that the law seldom works in favour of the individual. There is a world of difference between what the law offers and what corporate lawyers make possible. The individual's 'strong fair use defense ... is of little practical importance due to the procedural crippling of the general public's access to fair use.' Nonetheless, the legal doctrine of fair use 'establishes YouTube as something other than a community of pirates.'[39]

YouTube Celebrities

Along with the YouTube corporation itself, other types of member within this community are media celebrities and their parent corporations. Janneke Brouwers notes that when celebrities such as Oprah Winfrey become members of YouTube, 'they stand under close scrutiny of their fellow members, who can freely express their opinions about the presence of corporate identities in relation to their convictions of what YouTube is and should be.'[40] The presence of celebrities within YouTube has created an ongoing debate about its proper use by media corporations. Some embrace their presence and others see the entrance of celebrities and corporations as the end of a golden age for YouTube.

Oprah's second YouTube video, *Oprah's Message to YouTube* (2,372,074 views), provides an indication of how YouTubers struggle to define the proper use of YouTube and who should or should not be allowed membership. Many comments attached to the video demonstrate that she is seen by numerous YouTubers as an outsider who does not belong in the community. YouTubers can vote a 'thumbs up' or a 'thumbs down' on any comment attached to a video, which allows us to interrogate the way opinions are received. The majority of early comments that approve of Oprah have a negative rating.[41] Members see YouTube as a space for amateur use only (even though it obviously is not). Comments on the video such as 'Get off YouTube Oprah, no one wants you here,' 'YouTube has nothing to do with you, now pack up your billions and go home,' 'Youtube is for people who AREN'T already on television. Give someone ELSE a chance,' 'please stay on regular TV Oprah!' and 'Don't you think you have enough exposure! Get off YouTube!' receive high levels of 'thumbs up' ratings. Comments such as 'Hi Oprah, welcome to the Tube' and 'Oprah is cool I love her' receive 'thumbs down' ratings.

Some amateur videos that are critical of Oprah receive higher viewer counts than many of Oprah's own videos. In 2008 Oprah's videos generated an average of approximately 50,000 views. Videos such as *The Church of Oprah Exposed* (over 8 million views) and *Oprah Denies Christ* (over 4 million views) demonstrate that, like the rest of the Internet, YouTube is a dangerous place for brands. Curiously, when media professionals use blogs as a promotional tool, their professional identities 'are conserved and protected, meaning that any information that could potentially damage television brands or careers is kept firmly off-stage.' YouTube does not operate in the same manner as a Web-based blog and thus does not afford the same level of image control by celebrities and their handlers. Whereas blogging by media professionals 'actually reinforces the cultural and symbolic power of media producers,' YouTube provides a platform that may indeed undermine the cultural power of corporate media producers.[42]

Oprah fans argue with her detractors in written comments and videos (*people still love Oprah*) and even write love songs for her (*Oprah We Love You*). Other amateur videos question the validity of Oprah's participation on YouTube. We cannot generalize from the reaction to Oprah as to whether or not the majority of YouTubers are against celebrities' using YouTube. We can say with confidence that YouTube is a place where identities are contested and communities clash. Media corporations are both reinforced and undermined by their various practices of online expression.

Throughout the 1990s Internet users feared that corporations would take over the Net. The same fear can be seen within the YouTube community. Some argue that the golden age of YouTube is over (*RIP the golden age of YouTube*), while others welcome celebrities and also use it as a place to promote video clips of their favourite stars. Brouwers notes that any one claim about YouTube's proper use and identity will almost inevitably be challenged. What we are witnessing in the course of these community debates is an agenda-setting process that determines 'what part of YouTube's identity is open for debate and negotiation.'[43] In the Oprah debate the main issue is YouTube's status as a place that some think should be for amateurs only.

There is no one authoritative YouTube identity, but there is one dominant YouTube community – the community of amateur videographers. Their numbers will most likely always exceed those of participating celebrities and media corporations. Thus, YouTube's identity as a special place for amateurs will continue to play a significant role in how the audience receives and interprets celebrities and content from the corporate world. As Brouwers observes of the Oprah debate, 'Through their text comments and video responses YouTube users define their platform as a creative alternative to television, a place for user-generated content … In many instances Oprah is seen as a threat to YouTube's identity, an infiltration or corporatism and of a "foreign" medium, and faced with this threat, the otherwise fragmented YouTube unites to defend their position on YouTube.'[44] Amateurs see themselves as the authentic members of the community.

As others have noted, YouTube's corporate partners are seen as exploiting the community.[45] By entering the domain of YouTube, media celebrities and their parent corporations are fostering collaboration with and resistance to television's domination of viewing habits. Perhaps in the future commercial content will dominate YouTube's audience. Yet the very presence of corporate personalities and content within YouTube fosters a sense of YouTube as something different from television.

Productive Fan Communities

The phrase 'fan community' occurs over 8 million times on the Internet and 'fan video' occurs over 156,000 times on YouTube. Throw a stone into cyberspace and chances are good that you will hit a fan of some television show or other media product. Fans have moved from the edges of culture and are now accepted as a significant and serious aspect of society. What constitutes a fan is open to debate, but perhaps the most straightforward way to think about fandom is in terms of intensity. I watched all of the Buffy and Angel television series but remain emotionally disconnected from these shows – not a fan. I have followed most of the various Star Trek series since 1966, seen all the movies, cried when Spock died, read many of the books, and own my own phaser. I am a fan of Star Trek. Nonetheless, I have never gone to a Star Trek convention, joined a club, or otherwise socialized with a community of Star Trek fans. Sometimes the experience of being a fan is a private affair and sometimes it is deeply social. While there is much that has been written about fan culture and the Internet, here I will focus on how some fan communities create videos that critique media personalities. In the following case Britney Spears provides a slow-moving target.

One of the most famous fan videos on YouTube features the amateur-professional performer Chris Crocker crying hysterically over YouTubers' ridicule of Britney Spears's 2007 MTV Music Awards performance. Her performance was called 'awkward and apathetic' by the *New York Times* arts and culture writer, Dave Itzkoff.[46] Crocker's video *Chris Crocker – LEAVE BRITNEY ALONE!* (September 2007) shows Crocker in tears and pleading with the YouTube community to stop criticizing Spears. The video and its various copies have been seen over 30 million times. The original video was viewed over 26 million times and garnered 394,181 comments (one of the highest numbers of comments attached to any YouTube video). The majority of the comments express negative attitudes towards Crocker, homosexuals, and Spears.

Within the comments can be seen a debate over what a true fan is: 'Don't mind the haters, Chris. You continue to show your love and support and the rest of her true fans will, too,' and 'I absolutely love what you're doing for her. I think you're a true fan. Just because someone is going through a hard time doesn't mean everyone needs to be bashing.'[47] Others accuse Chris of faking and trying to become a celebrity himself. After the video Crocker was invited to appear on various national television talk shows. Fan videos begat fame.

Crocker's tearful performance was extensively parodied by other YouTubers. One parody created a mashup of various SpongeBob Squarepants videos and the soundtrack from the Crocker video, in which SpongeBob pleaded for Britney to be left alone. This in turn led to a

A parody of Chris Crocker's famous YouTube meltdown by one of my students.

clever appeal on SpongeBob's behalf (*Leave Spongebob Alone!*) which was seen a remarkable 811,041 times. Fan communities overlap within YouTube, which often leads to intra-community dialogue via videos and comments. A Star Wars fan used the image of Darth Vader to create the video *Darth Vader Command You to "Leave Britney Alone"!* and another donned a storm trooper helmet and recorded a commentary on the original Crocker video (*Re: LEAVE BRITNEY ALONE!*). In other amateur YouTube videos we see puppets, Adolf Hitler, and Michael Jackson pleading for Britney Spears to be left alone. Even a seven-year-old girl re-enacts Crocker's performance (*Leave Britney Alone – 7 year old girl*). Star Wars, SpongeBob, and Spears fans intersect in a complex web of parodies and inside jokes.[48]

Crocker's original performance has also led to the creation of a genre of 'leave alone' videos in which videographers parody his performance and plead on behalf of various personalities and subjects in videos such as *Leave General Petraeus Alone!*, *Leave George Bush Alone!!*, *Leave Indians Alone*, *LEAVE DORA ALONE!!!*, and *Leave Barney Alone!!!!*. There is even a *Leave Strangelove Alone!!!!* parody. Here we see how the storytelling on YouTube creates its own self-referential, reflexive culture. This amateur performance culture appropriates corporate content and themes and generates media genres and fan communities that are indigenous to YouTube.

As fans, YouTubers also participate in the meta-community of YouTube itself. They comment on each other's amateur videography and subject the artistic work of fellow amateurs to the same critical analysis that targeted Spears. When a Star Wars fan dons the costume of Darth Vader to parody the song of another YouTuber, we witness the complex interchange of meanings that can occur between YouTube's communities.

Fans are a definitive example of an emerging participatory culture within our media-soaked society. The 'Leave Britney Alone' corpus of amateur videos demonstrates how various fan communities interpret media events through the lens of their own fan cultures. The online audience used fandom's media products to create video commentaries on Britney Spears and thereby participated in the spectacle of a celebrity's lacklustre performance. YouTube enables communities of shared interests to interact through words and moving images. YouTube can also turn fans into celebrities.

Adam Nyerere Bahner wrote and performed the song *Chocolate Rain* under the pseudonym Tay Zonday and uploaded it to YouTube on 22 April 2007. Within a short time his song was covered by professional musicians and widely parodied on YouTube. Throughout 2007 Bahner made occasional appearances on national television. He was featured on the front page of the *Los Angeles Times*, was covered in *People* magazine, and appeared on CNN. By the fall of 2008 *Chocolate Rain* had gained over 30 million views on YouTube.

Bahner's is one of many such cases in which we witness a musician gaining national media attention and minor celebrity status as a result of his amateur music video. YouTube and amateur music videos are a deinstitutionalized route to such recognition, but what happens to a YouTube member when he or she gains celebrity status? Often, the intent of the YouTube luminary is to translate Internet fame into television, movie, and recording contracts. We usually (but not always) witness the amateur seeking to translate the new status into cash by integrating himself into the corporate entertainment sector as yet one more product. In this case Bahner remixed his song and, with guest vocals supplied by rapper Mista Johnson, made the new version, *Cherry Chocolate Rain*, as a viral ad for Cherry Chocolate Diet Dr. Pepper.

Whenever a YouTuber achieves celebrity status, he or she is bound to be parodied by other members of the YouTube community. *Chad Vader* is a fan film series and a YouTube personality created by Aaron Yonda and Matt Sloan. Yonda and Sloan are media professionals, but their parody *Chocolate Rain by Chad Vader* remains in the territory of fan videos. This parody refers to the original song, the Star Wars franchise, and the *Chad Vader* film series and demonstrates the high level of interpretative skills found within the YouTube audience, which is continually engaged with intensely intertextual media texts. Other fan parodies have used the Cookie Monster from *Sesame Street*, Alvin and the Chipmunks, and the animated video game characters Sonic, Rurouni Kenshin, Raving Rabbits, and Mario. In her analysis of YouTube fan videos Nicolle Lamerichs suggests that, 'by mocking the culture within YouTube,' fan parodies make YouTube's culture explicit.[49] Fan communities and their participatory culture show us what YouTube is all about.

In the middle of the last century the mass audience was largely conceived of as passive – a couch potato worshipping at the altar of the 'boob tube.' With the advent of digital technologies and new theories about the active character of the audience, we now see the viewer as interacting, participating, collaborating, and co-producing. Henry Jenkins suggests that the 'greatest changes' within media culture are 'occurring within consumption communities. The biggest change may be the shift from individualized and personalized media consumption toward consumption as a networked practice.'[50] Jenkins wrote these words in his book *Convergence Culture: Where Old and New Media Collide* (2006) just as YouTube and mass involvement in amateur video production were kicking into high gear.

Jenkins's notion of participatory culture positions the individual as a consumer. We exist in 'consumption communities' and consume within networks. It is certainly true that the audience consumes media and goods. Jenkins is also correct in suggesting that 'we may have greater collective bargaining power if we form consumption communities.' Certainly, the YouTube community is able to exert pressure on its corporate landlord, however lightly.

YouTube and amateur video practices highlight how participatory culture also is driven by our new position within productive communities.

The YouTube audience exists in production communities. As Angelina I. Karpovich notes, 'the rate of production of fan videos has increased exponentially ... the current rate of amateur fan videos matches, if not surpasses, the rate of professional music videos.'[51] The rate of amateur video production in other areas far outpaces the rate of professional production. Along with Jenkins's notion of consumption communities, we need to heed our emerging role in networked productive communities. Nancy K. Baym's suggestion that 'fandom is a harbinger of cultural phenomena to come' applies to the entire YouTube community of amateur videographers.[52] Amateur video is a harbinger of new forms of productive media practices.

Haters, Spammers, and Other Deviants

A community is made up of those who fit and those who do not fit well with its norms and tacit rules. Internet communities have long been plagued by those who do not play well with others. One such type of person is known as a 'hater' – a person who posts rude and often racist, sexist, homophobic, or obscene messages. The word *hater* appears over 111,000 times on YouTube. The problem of haters itself is the subject of a subgenre of YouTube videos. These videos tell us much about the YouTube communities' tacit norms and how amateur videographers respond to bad behaviour on the 'Tube.

Some of the videos that address the issue of haters draw large audiences: *A Message To All Haters!* (5,745,229 views), *Kevjumba Responds to Haters* (1,294,408 views), *BAN SARCASM FROM YOUTUBE!!!* (1,898,873 views), *TILA TEQUILA'S MESSAGE TO HATERS,* (330,562 views), *To My Haters, Flaggers & Trolls* (37,275 views). Both amateur and professional videographers create videos that respond to their haters. Not all YouTubers see haters as a problem. In *YOUTUBE HATERS ..!!* ThePeacefulCat writes of his experience: 'You either love them or hate them! Or even both. My haters in the past have not been so bad. They only stuck around for a few days before moving on to someone else.' YouTube videographers demonstrate a high level of awareness of the haters in their midst and often encourage each other to ignore the haters. In *Re: Haircut Experience update!!!* NorCaLg33k writes, 'Dave, buddy keep up the videos and don't let the ignorant haters get you down.' Some YouTubers make videos directly addressing haters, and haters also make video responses. Even hater videos can be highly creative.[53] Not all YouTubers define hating behaviour in the same way, nor do they react to it in the same manner.

Email discussion groups, along with blogs and social network sites, are renowned for the amount of hatred that can be found within them. It is quite possible that this reflects the cultural conditions of the real world. In *The Argument Culture: Stopping America's War of Words* Deborah Tannen finds that American public discourse is based on the metaphor of war. Americans live in 'an atmosphere of unrelenting contention.'[54] Anthropologist Peter Wood describes a 'national epidemic of anger' that has infected American society, which he calls the New Anger. This New Anger is found throughout the United States because it is 'the expression of a new cultural ideal that emphasizes the importance of individual authenticity achieved through the projection of personal power over others.' Americans have moved 'beyond vituperation to a kind of anger that luxuriate[s] in its own vehemence ... New Anger is about declaring one's identity as it is about taking umbrage at someone else's infraction.'[55] This public expression of anger may have found a natural home on the Internet.

The question arises: is the angry character of YouTube related in some way to the cultural context of its homeland – America? It would be foolish to blame America for all the haters on the Internet, but, at least for now, the country constitutes the largest population on YouTube. Hate speech plays a prominent role in American discourse and receives special protection under U.S. laws. As Cass Sunstein told readers of the *New York Times*, 'The U.S. is pretty unusual [in] providing [the] broad protection we do to hate speech.'[56]

Hate messages have been cited in the past as a problem in Google's online communities, and it is not difficult to find toxic speech on YouTube, such as the comment 'I believe that a good Israeli is a dead one.'[57] It is not uncommon to see racist videos such as *The problem in this country is niggers* posted to YouTube and quickly removed. Others, such as *I Hate Black People Because ...* (517,541 views), have been on YouTube since November 2006.[58] Videos within YouTube are littered with racist, sexist, misogynist, homophobic, anti-Semitic, and obscene hate speech. Also, there is a great deal of violence in YouTube's offerings.[59] YouTube is in danger of turning into WeHate. Fortunately, many YouTube amateur videographers also make videos that expose hate speech, contest claims of hate speech accusations, and otherwise bring to light the problem of hate speech.

Perhaps, as *Le Monde* once famously suggested, we are all Americans now – we all suffer from too much contention and anger.[60] Or perhaps the hatred we see on YouTube is due to the condition of online anonymity. Patricia G. Lange observes that some scholars claim that online hostility 'results from assumed online anonymity rather than from social or culture dynamics which may occur offline.' Yet we cannot isolate on-line behaviour from the influences of its offline cultural and psycho-social contexts. Lange's own research demonstrates that, contrary to earlier expectations, 'the addition of facial and bodily information in video does not guarantee cordial interaction.'[61] Video may actually in-tensify the expression of anger and hatred.

Two other categories of problem behaviour on YouTube are 'spam-mers' and 'flaggers.' Flaggers falsely flag a video for violating YouTube's content regulations. Flagging a video can lead to YouTube's removing the video or deleting a member's account. It is a common censorship strategy within YouTube and highlights the corporation's uneven appli-cation of its content policies. Minke Kampman describes the flagging process as resembling 'a governmental democracy in which YouTube forms the government and flagging videos are our everyday miniature referenda.'[62] Flagging is commonly used against gay-themed content by homophobic individuals. This tactic is known as 'fagging.' Members of the gay community respond with videos such as *Flagged or Fagged* (5,158 views) that point out the improper nature of such censorship. Kampman correctly notes that the flagging feature erodes the democratic nature of YouTube, as in the end neither YouTube nor flaggers are transparent and accountable in their actions.

Spammers (also called 'spamtards') use YouTube to send unsolicited advertising to YouTube members or otherwise abuse the system. The YouTube community has made numerous videos complaining about abusive behaviour and explaining how to take action against flaggers and spammers. Other videos explain how to spam YouTube accounts. Software engineers have been playing a cat and mouse game with email spammers for two decades and are just beginning to develop methods for automatically detecting video spam.[63] Haters, flaggers, and spam-mers are unlikely to spell doom for the YouTube community – not when there are so many involved in various neighbourhood watch programs.

A Highly Reflexive Community

Whether fighting unfair copyright actions, debating the purpose of YouTube, or grappling with anti-social behaviour, the YouTube com-munity proves highly reflexive. There is much reflection on the norms and ideals of the community and also a constant monitoring of other people's behaviour. According to one study, at least 10 per cent of highly popular videos were 'explicitly concerned with YouTube itself.'[64] This is a community that puts considerable energy into discussing the community itself.

As a brand, YouTube has a deeply involved community of users. Two of the more common genres found in this productive community are video diaries (also called video blogs) and parodies. Both modes of cultural production are highly interactive and prone to crossing boundaries. Although no one definition of the community is authoritative, amateurs who play by the rules claim the high moral ground, celebrities are regarded with suspicion or outright disdain, and haters and spammers receive near universal condemnation. Burgess and Green suggest that these debates and complaints reveal an implicit social contract that structures the participation of amateur videographers but 'is only made explicit once it appears to be broken.' Echoing Mary Douglas's notion of communities, they suggest that 'what is at issue for all these players in these controversies is the extent to which they have an influence on the future of the community in which they have so much invested.'[65] Commercial media, haters, spammers, and the village cop (YouTube) all represent forces that could change the character of the community, and all are subject to monitoring and debate.

Even though there are over 300,000 videos uploaded to YouTube every day, not every member participates to the same degree. By one estimate only 1 per cent of YouTubers contribute videos.[66] Other estimates range between 2 per cent and 10 per cent. Only a small fraction of the community is involved in video-based interactions (video replies), and most video replies are directed at a small fraction of YouTubers. While it is not uncommon to see a flame war erupt within written comments, 99 per cent of video replies involve interactions in which each user creates only one video reply.[67] In other words, it is rare for conversations within YouTube to persist over a series of multiple video replies. Given the rapidly changing character of consumer video technology and the relative novelty of mass participation in online video-making, we should not infer too much about the future of video-based dialogue from these early statistics. It may yet be the case that, in the near future, video-based dialogue becomes far more common.

Why and How Do You 'Tube?

As recently as 2008 John C. Paolillo observed that 'there is no clear picture of how people use YouTube and why.'[68] Why do people participate in YouTube? The surrounding culture of confessional practices can be said to predispose individuals to make video diaries about their lives. The many styles of representational practices such as film-based home movies, reality-TV shows and audience-produced television material (*America's Funniest Home Videos*) also provide us with models for self-representation. In other words, western society encourages individuals to seek out media forums for self-expression.

We are socialized to embrace self-expression via video. Advertisements for consumer electronics tell us this is normal, television shows us how to do it, our peers (and our children) lead the way, and social norms reassure us that it is acceptable. It is also fun to be a YouTuber. Pleasure and vanity are great motivators. Certainly, the high valuation placed on fame and being on television must also compel some to seek out an audience on YouTube. Thus, we can speak of pervasive cultural and psychological forces that predispose us to participate in amateur online videography.

There are also more material matters of utility and convenience that explain our YouTube compulsions. People use YouTube for distributing video holiday greetings, wishing friends happy birthday, celebrating anniversaries, births, and deaths. The answer to the question 'why do you 'Tube?' is as broad as the answer to the question 'why do humans communicate?'

To some, YouTube remains a mystery. As Lange observes, 'People who do not regularly participate on YouTube may not understand why people watch seemingly poor quality or odd videos.' Certainly, the picture of how and why people use YouTube becomes a little clearer when one is involved in the community. I have been a member since August 2006 and have watched thousands of YouTube videos. Since then, I have created and uploaded hundreds of YouTube videos, some of which remain on my channels and some of which I have deleted. I have also contributed written and video replies to YouTube. I have used YouTube to communicate with an online gaming community, distant relatives and friends, colleagues, students, and the YouTube audience at large. As Lange points out, amateur video production serves important social functions, 'creating and circulating video effectively enacts social relationships between those who make and those who view videos.'[69] Amateur video practices also can arise from the simple desire to play, which itself can be a highly social activity.

One of my amateur videos is a staged scene created with the help of students at the University of Ottawa. In the fall of 2008 I asked students to bring in old cell phones that no longer worked. I selected one of these phones and then coached a student to pretend to answer her cell phone while I was lecturing. The result is *Dr. Strangelove Smashes a Cell Phone*, a parody of the angry professor genre of videos (*angry professor*). My inspiration for making this video was a joke played on students by a French professor (*Prof smashes laptop: what REALLY happened*). I also wanted to see what happens when the same incident is recorded by multiple cameras, so I invited the class to film the action and put their videos up on YouTube. The result is a series of videos on YouTube that capture the event from different angles and show different editing styles and qualities of video reproduction.

Whether it is teenagers recording each other playing pranks or professors hamming it up in the classroom, here we encounter a central cultural characteristic of amateur videography and its YouTube communities – fun. As Jean Burgess and Joshua Green have noted, playfulness and affect 'characterize the "common culture" of YouTube.'[70] We do not just watch YouTube. We like to play with YouTube and its audiences.

My approach to the study of YouTube follows the practice of autoethnography that is commonly found within media and cultural studies. I make no special claims for a mode of study that involves deep immersion into the thing studied. As Matt Hill points out, autoethnography can degrade into narcissism (which he finds Fiske guilty of) or an overly optimistic evaluation of the audience's ability to provide a full self-explanation (which he finds Jenkins guilty of).[71] As the epistemological problems associated with ethnography are well documented elsewhere and perhaps are as irresolvable as any other method, I will not dwell on these issues here.[72]

Hill's accusation that fan ethnographies assume that both fans and academics can fully explain their cultural practices does not apply here. In watching YouTube we are continually confronted with what we do not yet know – the perhaps, however, and maybe that continually qualify the few certainties of the 'Tube. Oddly enough, Hill's highly sceptical assessment of autoethnography leads him to employ psychoanalysis as a method of inquiry. This strikes me as jumping from the frying pan into the Oedipal fire. Ethnography and autoethnography do not do away with the problem of authority when they represent other peoples' representations. It does not solve the 'current crisis in representation' within academia, but it does have the benefit of treating representations as social facts.[73]

There is a subgenre of amateur videos on YouTube that explores the question 'why do I YouTube?' Andsanp created his own YouTube channel once he discovered that people on YouTube were sharing information about themselves. Andsanp went on to make ninety YouTube videos over the next year and a half. Someone saw one of his early videos and encouraged him to continue with video making; in *Why I YouTube* he stated, 'that's when I started to realize that this world has some good people in it … you go on YouTube and you see people encouraging each other and helping each other and filled with compassion for each other … there is a lot of love and it makes you feel good so that is why I YouTube.' In early 2008 it appeared as if Andsanp had left YouTube, which prompted Caryn, one of his YouTube friends to write, in *andsanp is gone?* 'You are one of the positive forces here and without you YouTube will not be the same … I have also had some very good experiences and met a few wonderful people like yourself that have made this whole experience enjoyable for me. I will never forget the wonderful people that I have met during my time here.' Clearly, people are creating social networks and emotional bonds within YouTube.

Crissy, a twenty-six-year-old female who has made ninety-seven YouTube videos, explains that, when she first saw people talking on YouTube, she thought, 'I can do that.' Crissy's video *RE: Why I youtube* explains that she likes YouTube because people can watch her grow on YouTube. She reflects on the fact that initially she was shy but now she boldly presents herself and her opinions. She likes YouTube because of all the friends she has made there: 'I don't even watch TV anymore, YouTube is my TV. Everything on here is just great. The community, the friends, the people, there are haters but you just label them as haters and you move on but your friends, your true friends are on here. You get to know people. They show you a side of themselves that they don't show their close friends … here people accept you for who you are regardless of how weird or crazy or strange you are they "get you" and I like that a lot of people are getting me … You guys make me feel good about what I do and it makes me want to do more [videos].' While the explanations are many and varied, one that is repeated throughout YouTube is the experience of encountering others and making friends. Amateur online video practices bring strangers together and often turn them into friends.

Although the 'why' of amateur video practices are fairly transparent, the technical and social structure of YouTube itself remains somewhat opaque. Various macro-level analyses of large-scale data sets have been conducted, and there is a growing body of micro-level analyses of YouTube communities. Emerging from these early studies is a picture of YouTube's technical and social structure. The following is a series of observations gleaned from several early macro-level analyses of YouTube traffic. The results allow for only some broad generalizations but nonetheless provide interesting food for thought.

Within YouTube there is a core-periphery social structure that mirrors other social networks. A small minority of users produce videos that provide thematic content (such as anime or punk music) around which socially connected groups form.[74] The most popular content themes are music (22.9 per cent), entertainment (17.8 per cent), and comedy (12.1 per cent). As of mid-2007 almost all YouTube videos (99.1 per cent) were no more than 700 seconds long, owing to the ten-minute limit imposed by YouTube on most members. The vast majority of videos (98.8 per cent) were less than 30 megabytes in size with an average video file size of 8.4 megabytes. In 2007 there were approximately 42.5 million YouTube videos, which required 357 terabytes in disk space.[75] By mid-2009 there were over 95 million videos on YouTube.

Researchers have found that a video's growth in popularity slows down as time passes. Videos have a short active life span – they are watched frequently in a short span of time and then are almost never watched.[76] This does not apply to many of the most popular videos, as they continue to attract viewers because of their high visibility within YouTube's ranking systems and because of word of mouth, email, and blogs. Viewing choices appear to be somewhat insensitive to the age of a video but 'videos that are requested the most on any given day seem to be recent ones … Over a one-day period, roughly 50% of the top twenty videos are recent. However, as the time-window increases, the median age shifts towards older videos. This suggests ephemeral popularity of young videos.'[77]

A more recent study by Brian M. Landry and Mark Guzdial concludes that popular amateur videos on YouTube tend to show individuals engaged in uncommon activities. These highly popular videos average under six minutes in length (89 per cent), which suggests that there is a relatively short attention span among the YouTube audience. Landry and Guzdial find that 60 per cent of these popular videos are 'reminiscent of mainstream variety television shows that showcased everyday people performing not so everyday acts.' YouTubers like to watch ordinary people doing extraordinary things. The second-largest genre of popular amateur videos (36 per cent) concerned activism and outreach. A campaign for free hugs, an invitation to explain why a person uses YouTube, and other requests for involvement make up this popular category. Instructional videos are a distant third at 3 per cent. Landry and Guzdial also find that 90 per cent of popular videos are not plot driven, but 89 per cent do contain emotional content. Less than half of the popular amateur videos contain a soundtrack (48 per cent) and 29 per cent use sound effects. Following the format of television shows and movies, 41 per cent of videos have an intro sequence and 49 per cent include an outro sequence (outros are credits and similar acknowledgments).[78]

There are hundreds of videos on YouTube that explain how to make videos. These instructional videos partially explain the tendency to copy two of televisions more popular formats – the confessional story (as seen in video diaries) and performances of an extraordinary nature. Amateur videographers are also incorporating some of the commonly seen elements of professional products – soundtracks, sound effects, intros and outros. In common with the much earlier practice of film-based home movies, YouTube has a growing body of how-to discourse. Miles Dyer, a twenty-one-year-old student at Hertfordshire University in England, made the video *How to make your 1st Vlog! START TODAY!* (420,475 views). This seven-minute instructional video reviews basic techniques of video blogging and garnered 434 video responses and over 3,000 written comments. Dyer's channel attracted 21,484 subscribers. This is another example of the extremely high level of interaction that occurs within YouTube.

The most popular content on YouTube is made by professionals and media corporations, yet amateur content can also attract over 100 million views.[79] Within a day or two of uploading, 90 per cent of new amateur YouTube videos are watched at least once, and '40% are watched over ten times.' The initial response to a video is a good predictor of its long-term popularity: 'if a video did not get enough requests during its first days, then, it is unlikely that [it] will get many requests in the future.' Nonetheless, in rare instances an old, overlooked video can become popular. The more popular a video is, the more likely it is that we will find multiple copies of the video on YouTube. Curiously, over 80 per cent of these copies are from one-time uploaders (people who upload a video to YouTube only once). Also of note, the rate of deletion of videos by the uploader and by YouTube combined is only 0.4 per cent and only 5 per cent of these deleted videos violated copyright laws. Furthermore 'illegal uploads are more common amongst highly ranked videos.'[80] This also suggests that the audience is highly attracted to commercial media and that only a small percentage of copyright violations on YouTube lead to a video's being deleted.

The single biggest technical issue facing YouTube is known as 'scalability.' In 2007 YouTube accounted for 10 per cent of all traffic on the

Internet. Given that only a small percentage of YouTube users upload videos, if this figure is increased significantly, it might spell disaster for the entire Internet. As one group of researchers note, YouTube has a significant impact on Internet traffic, 'and itself is suffering from severe scalability constraints.'[81] The revenue-generating potential of YouTube and the small army of network software engineers who are addressing the issue probably ensure that the Internet and YouTube will continue to overcome the technical hurdles that lie ahead.[82] There is simply too much money at stake. In 2008 one-third of American broadband Internet users chose to pay more for faster connections.[83] When it comes to getting more at a faster rate, Internet users are willing to pay. This suggests that mass involvement in amateur online video will prove quite lucrative for the telecommunications sector.

Ordinary People and Their Extraordinary Videos

YouTube and amateur online video provide a new window into the extraordinary nature of everyday people. The practices of YouTube's amateur videographers are similar to those of very early moviemakers. Cinema prior to 1906 has been described by Tom Gunning as a cinema of attractions – an exhibitionist cinema that directly acknowledged viewers and invited them to look. Gunning describes the cinema of attraction as 'the direct address of the audience, in which an attraction is offered to the spectator by a cinema showman.'[84] These early amateur films drew attention to themselves, whereas the current style of professional cinema associated with Hollywood encourages the viewer to forget the presence of the producer and the screen and be absorbed into the unfolding story.

One of the main purposes of YouTube videos is to be seen – to draw an audience. Teresa Rizzo suggests that YouTube functions as a cinema of attractions by engaging an audience that is 'highly attuned to attractions.' This is not merely the same audience phenomenon of a century ago. A media- and experience-saturated society has created an audience that is involved in an intense search for spectacles that could occur 'anywhere at anytime ... exhibition, shock and sensationalism become everyday events ... Early cinema was an event in itself that required audiences to put time aside to attend the theatre. YouTube attractions are available all day and night.'[85] We used to go to see spectacles in dark theatres in the company of strangers. Now a constant stream of YouTube spectacles rushes towards us wherever we may be.

In common with many other Internet users, much of my YouTube viewing occurs as a result of a clip arriving in my email box or on my Facebook page having been sent by a friend. YouTube attractions operate like the cinema of attractions by demanding our attention, stimulating curiosity, and delivering pleasure through an endless supply of spectacles.[86] Intriguingly, Rizzo also suggests that 'the dialogic nature of video sharing sites produces a kind of hyper-attraction as not only are subscribers invited to watch clips but post comments or respond through video posts.'[87] We watch the spectacle, comment on the spectacle, forward the spectacle to our friends, and sometimes imitate or parody the spectacle.

In what follows I will survey some of the extraordinary spectacles of YouTube's amateur videographers and explore this cinema of attractions that millions of Americans return to each day. Being a cat-lover, I'll begin this journey with one of my favourite YouTube videos, *An Engineer's Guide to Cats* (3,806,513 views), written and directed by Paul Klusman in October 2008. This seven-minute video delivers a deadpan style of humour and gives an overview of life with three cats. Klusman relies on a written script and simple editing to tell his story. His careful planning, scripting, acting, and use of multiple locations is relatively uncommon on YouTube. Like all highly popular YouTube videos, this one has been copied to numerous other Internet video sites, personal blogs, and subject-oriented blogs (in this case, of course, pet, science, and engineering blogs). Video responders to this work include one YouTuber who wears a Storm Trooper's helmet (apparently a rather common sight on YouTube). There are no obscene remarks within the video's 7,600 comments.

The often repeated claim that online video 'is not succeeding in telling stories' is contradicted by the storytelling that takes place in online video diaries and by more complex works such as Klusman's.[88] When every picture tells a story, it is odd to see claims about the lack of story in even the most incoherent amateur video. There are millions of stories on YouTube, but they do not appear in the formulaic narrative moulds provided by television and movies.

The notion of storytelling has been clouded by overly simplistic definitions such as 'A story is a structure that encompasses a beginning, a middle, and an end,' and 'A story has three basic parts called the beginning (exposition), the middle (conflict), and the end (resolution).'[89] We have always constructed stories from fragments and co-authored stories by combining, with our own imagination, what we see and hear. Any story only partially resides in the text (or video) and is co-produced in our acts of reading and viewing. The answer to the question 'what is a story?' remains unresolved and will be challenged once again by amateur video practices.[90]

The popularity of their videos often takes amateur videographers by surprise. Klusman, who bears a striking resemblance to George Clooney, relates that his video resulted in numerous marriage proposals. He also offers T-shirts for sale that are based on the video. One of the first things amateur videographers do when their YouTube videos become famous is to sell T-shirts. Another pattern found here is that many amateur YouTubers who do make a popular video are unlikely to repeat the success. Whereas *Cats* attracted an audience of over 3 million, Klusman's subsequent videos have drawn under 20,000 viewers (certainly an admirable number, but not the stuff of YouTube fame).

Cats are a popular subject on YouTube, where many cat videos draw millions of viewers. One amateur video of a cat playing with the keys of a piano garnered over 14 million views (*"NORA: Practice Makes Purr-fect"* – *Check the sequel too.*). Another video of two cats meowing has an astonishing 21 million views (*The two talking cats*). However, few, if any, YouTube cat videos match the compositional complexity of Klusman's video. Most cat and pet videos are unedited scenes displaying unusual behaviour in domestic settings, often in the form of a single, continuous shot using a hand-held camera.

Video titles play an important role in attracting audiences to amateur work and in framing the meaning of the video. The twenty-three-second video *Sasha CHEER for Team U.S.A!!!* depicts a small dog sitting on a kitchen floor and howling each time a woman says, 'Sasha cheer for team U.S.A.' Viewed over 20 million times by the end of 2008, the clip was the number-one viewed video ('All Time') in the YouTube categories of Canada, Sports – Canada, and Sports. The commentary attached to the video is filled with anti-American comments and a nationalistic debate between Canadians and Americans. Here the title alerts YouTubers to the sports and nationalism themes and the comments reflect these themes.

Another popular category of YouTube videos involves children saying funny things. In one video a mother asks her three-year-old daughter what would happen if a monster came into the house. The child replies, 'I'm going to kick his ass.' Seen over 14 million times, *He's gonna kick MY ass?* is also an unedited, single, continuous shot from a hand-held camera in a domestic setting. A common feature in this genre, we hear the parent speaking from behind the camera asking the child to repeat something that was just said and then correcting the child ('That is not a nice word. You should say kick his butt'). Again, here we find no parodies or obscene comments.

Highly popular videos often capture people engaged in everyday private behaviour of an innocent nature. Gary Brolsma, resident of Saddle Brook, New Jersey, in December 2004 made a grainy, low-resolution video of himself sitting at his desk in his bedroom lip-synching to the Romanian pop song *Dragostea din tie* and posted it to Newgrounds.com. Entitled *Numa Numa*, the video was subsequently copied to YouTube in August 2006 where it has been watched over 24 million times. Widely copied to other Web sites and viewed over 700 million times, *Numa Numa* is the second most watched viral video of all time (*Star Wars Kid* is the most watched video).[91] Perhaps the best thing to happen to O-Zone, the Moldovian pop group that produced the song, was Mr Brolsma. The video captures a moment of sheer pleasure and unassuming self-enjoyment. It shows how amateur video is transforming private moments of ecstasy into public displays that, like the earlier video, *Star Wars Kid*, can bring in their wake public humiliation.[92] As Alan Feuer and Jason George explained to readers of the *New York Times*, 'Even in the bathroom mirror, Mr. Brolsma's performance could only be described as earnest but painful.'[93] *Numa Numa* and the *Star Wars Kid*, the two most viewed videos on the Internet, together have garnered over 1.5 billion views.

Comments for the original *Numa Numa* video have been disabled, which suggests that they were less than polite. The video has been parodied hundreds and perhaps thousands of times in numerous languages. On YouTube alone there are over 31,000 references to the video and many parodies and remixes that attract millions of viewers. Lego fans are renowned for creating stop-motion animations of moments in popular culture and *Numa Numa* is no exception (*Numa Numa Lego*). The YouTuber Trooperman gets in on the action and dances to *Numa Numa* in a complete Storm Trooper costume (*Numa Numa Stormtrooper*). If there is something being parodied on YouTube, chances are good that we will encounter a Star Wars fan in full costume doing a parody.

Like other YouTube celebrities, Brolsma quickly turned his fame into a commercial venture in 2006 and promoted the sale of clothing, coffee mugs, and ringtones. He is featured in a new professionally produced version of the song on YouTube, *New Numa – The Return of Gary Brolsma!* (13 million views), but he does not appear to be active as a YouTube videographer. The 59,000 comments attached to his new video range between admiration and disdain. Some commentators express shock and disappointment that he so quickly tried to convert his fame into merchandising. In keeping with the YouTube community's central values, others find his subsequent commercial work lacking the spontaneity and sincerity of his first video.

Music videos attract the largest online audiences, and commercial music videos dominate annual listings of the most viewed YouTube videos. The Canadian princess of pop, Avril Lavigne, holds the record for YouTube's most popular music video, *Girlfriend*; at more than 121 million views, it is also the most viewed YouTube video in all categories. Avril's legions of teenage fans have been accused of gaming YouTube's rating system to inflate the view count.[94] Using Web-based systems known as auto-enhancers to inflate view counts within YouTube is common.

Hundreds of thousands of amateur musical performances compete for attention on YouTube. Jeong-Hyun Lim's rendition of *Canon Rock*, a version of Pachelbel's *Canon*, has been seen over 65 million times. Captured by a friend in the YouTube video *guitar*, the solo performance takes place in his bedroom. Lim, from South Korea, was subsequently mentioned on major media such as CNN and CBC. He is among tens of thousands of artists who share their work online and thus promote imitation and recirculation of art forms. Over 83 per cent of musicians and songwriters provide free music samples online.[95] As it does in so many other areas, YouTube enhances the circulation of performances and promotes innovation in artistic practices.

Academia has its own YouTube star in the person of Michael Wesch, assistant professor of cultural anthropology and digital ethnography at Kansas State University. Wesch, a visual anthropologist, is leading the way in defining the field of digital ethnography, was awarded the U.S. Professor of the Year Award in 2008, and has made numerous YouTube videos, two of which have attracted considerable attention. His first YouTube video, *Web 2.0 … The Machine is Us/ing Us*, (7.5 million views), explores the nature of digital text. He describes the video in the following terms:

> everything is connected. This is true on many levels. First, everything including the environment, technology, economy, social structure, politics, religion, art and more are all interconnected. As I tried to illustrate in the video, this means that a change in one area (such as the way we communicate) can have a profound effect on everything else, including family, love, and our sense of being itself … the ultimate promise of digital technology is that it might enable us to truly see one another once again and all the ways we are interconnected. It might help us create a truly global view that can spark the kind of empathy we need to create a better world for all of

Dr Michael Wesch, anthropologist of digital culture.

humankind. I'm not being overly utopian and naively saying that the Web will make this happen. In fact, if we don't understand our digital technology and its effects, it can actually make humans and human needs even more invisible than ever before. But the technology also creates a remarkable opportunity for us to make a profound difference in the world.

Inspired by Wesch, I made the video *The Real in the HyperReal (A Reply to: Web 2.0 ... The Machine is Us/ing Us)*. Whereas Wesch used writing and typing to explore the textual, hyperlinked character of the Internet, I used a series of videos embedded within videos to reflect the intense visual reflexivity of the self within YouTube. Mirroring the multiple redundancy of the holographic Web, the two-minute video repeats the question: 'So what happens to the media, what happens to us? When we become the medium. When we become the message. When we become the real inside the hyperreal. What happens to us?' I had in mind that Jean Baudrillard was only half right when he suggested that the real is lost within the hyperreal. We bring the real with us wherever we go. No matter how far we travel into the new domain of the hyperactive, hyper-real world of virtual reality, we are still confronted by the real of us, our communities, and ourselves.

Online amateur video practices are qualitatively different from the old film-based home movies of the mid-twentieth century because they are distributed over the Internet, not stuck in our living rooms. The distributed character of digital amateur movies like Wesch's enables us to interrogate the producer of the video and the conditions surrounding its production and reception. Both the background and the foreground to amateur digital movie practices are more transparent to us. In the case of Wesch, there is considerable background material on his motives for making the video and considerable foreground material for interrogating the audience's response. Of course, the Internet also allows us to query amateur videographers directly via email or video response.[96]

Wesch provides a perspective on the Internet that reflects anthropology's concern with how communication media shape expression: 'While no medium is completely restricted to specific types of expression, we might say that all media are biased towards certain types of expression due to their structure, format, and mode of creation and transmission.'[97] This notion of the bias of a communication, which also runs through media studies, is shared by anthropologists.[98]

In 1957 anthropologist Edmund Carpenter described how each medium has a bias that 'reveals and communicates a unique aspect of reality, of truth. Each offers a different perspective, a way of seeing an otherwise hidden dimension of reality. It's not a question of one reality being true, and others distortions. One allows us to see from here, another from there, a third from still another perspective.'[99] Each medium establishes different possibilities for expression. The networked, interconnected environment of online video is expanding our communicative possibilities. It is perhaps too early to proclaim the specific bias of online amateur video, but it may be the case that its bias is found in its social dimension. We do not merely watch online video. We engage each other in relationships through amateur online video practices.

There are thousands of bizarre and innovative media practices taking place on YouTube, most of which will escape our attention here, owing to lack of space. From teenagers filming each other vomiting to jackasses doing dangerous and harmful stunts, YouTube has more than enough to give any parent nightmares. Other amateur video practices are less disturbing. In what may be a genre unique to the Internet, individuals make video recordings of new purchases as they are taken out of their boxes. This amateur video practice is called 'unboxing.'

There are thousands of unboxing videos on YouTube, many of which include a demonstration and review of the product. The vast majority of these videos are made by young men and they almost exclusively feature technology, movies, music, and gaming products. This genre is attracting the attention of technology companies and is evolving into a new form of advertising. The videos can attract over 2 million viewers and have become so popular that companies have co-opted the genre and have made sponsored unboxing videos featuring their products (*Samsung Omnia (i900) Unboxing*).

I made an unboxing video called *Dr. Strangelove Unboxing an Easel*, which incorporates images of my Barbie art, still shots from the television show Dexter (which happened to be playing on the television at the time), and still shots of the easel being used to make a painting of the Ottawa skyline at night. The sole comment on my video thus far by a fellow YouTuber: 'This video sucks. You are one creepy bastard.' At the very least, the video is evocative.

Oddly enough, while YouTube restricts the representation of our own act of creating life (sexual intercourse) and our decision to take our own life (suicide), here on YouTube we can witness the delivery of life. Videos of hospital and home births are easily found on YouTube. Some are amateur video productions, such as *Unassisted Birth* (1,469,504 views), which documents the unassisted birth of Maelle Claire on 12 April 2007 on the bedroom floor of an American home. Only Maelle's mother, her husband Daniel, and her sister Jenny were present. Maelle's mother posted the video to her YouTube channel, socially skilled, and wrote extensively about the labour and birth experience on her LiveJournal site. Here we see how YouTube is only part of what is often an extensive trace of an individual's life across multiple Internet locations.

The video of Maelle's birth has been linked or copied to dozens of other websites around the world. Maelle appears in subsequent YouTube videos first as an infant and then as a young child. We see Maelle at seven months using sign language. Other videos show her playing with sister Kira and mom. Throughout these videos the domestic setting looms large in the background and we gain insight into the mundane architecture of the American family. Furniture, toys, pictures, pets (two cats), a Christmas tree, the kitchen table, exercise equipment, and bathtub toys reveal modern consumption habits and the American domestic lifestyle. The seasons pass and we witness Maelle's family and grandparents celebrating New Year's Eve around a dining room table. In the background the fridge doors are cluttered with the pictures, drawings, notes, and mementos that are a common fixture of household fridges – every home's central information server. The scenes are universal and as ordinary as white bread. In these snapshots of Maelle's family life we see bits and pieces of ourselves.

In one video we see Kira and Maelle in the back seat of the family car, which is stationary at a stop light. Their mother is filming them as she sits in the driver's seat waiting for the light to change. Once again, here we see a domestic amateur video being made in a car (a dangerous practice at the best of times). By the time I took my leave of Maelle, she was walking, talking, and singing. In the background was that universal fixture of modern life – the television.

There is within YouTube a community that supports and promotes the practice of unassisted births – home births that are not supported by medical professionals, doctors, midwives, or doulas. The risks associated with this practice are obvious, as the labour can be progressing normally, when suddenly problems arise. Fortunately, in Maelle's case nothing went wrong. Maelle, mom, and dad have embarked on an incredible journey together as they build their family. Maelle is one of the first of a new generation of children who are indigenous to cyberspace. These young representatives of Generation YouTube are born and will grow up on the 'Tube.

Often seen in the background of home videos that feature the youngest members of Generation YouTube is the ubiquitous television set. Yet their lives are not on television. Only with the passing of time will we know for certain what will happen to Generation YouTube as they live out their lives as members of a new virtual tribe. Generation YouTube is leaving a massive digital trail of words and images on the Internet, thereby changing the future by creating a new kind of collective memory, a new digital past.

Summary

Although virtual communities are often seen as inferior to real-life communities, in YouTube we find numerous communities that interact and extend into the realm of everyday existence. YouTube itself is a community member and gives special privileges to its paying corporate partners – media companies and their celebrities. Some YouTubers feel that YouTube should be for amateurs only and resent the use of YouTube by media celebrities. Nonetheless, numerous YouTubers also seek fame and fortune through their amateur video practices.

Amateur YouTubers are often their own worst enemies and can be seen abusing DMCA takedown notice procedures as they fight for cultural legitimacy and domination over other communities within YouTube. While in theory the law provides generous allowance for fair use of copyrighted commercial content by amateur videographers, legal practice and YouTube's actions tip the balance of power in the favour of media corporations. At the heart of YouTube is a struggle to free cultural production from the straitjacket of corporate control. YouTube's claims regarding its content policies are contradicted by its own actions (and failure to act) and by the often pornographic, violent, racist, hateful, sexist, homophobic, anti-Semitic, and misogynist content and commentary that can be found on its site.

One of the main cultural scripts among YouTube's army of amateur videographers is parody. Parody is a sign of fame and perhaps a necessary process for the creation of celebrity status and large audiences. Within YouTube fame is often accidental. As fame is a collective process, individual acts intended to draw and maintain large audiences often fail to generate the community co-production of content that leads to celebrity status.

Bad behaviour proliferates across YouTube's community. Haters, flaggers, and spammers demonstrate that there are implicit norms that guide participation in YouTube. As is true in any community, norms are violated, which leads to protests and debates. The community reflexively monitors the actions of other members and attempts to influence YouTube's content policies.

Finally, it takes a virtual village to raise a YouTube child. For all its deficiencies, YouTube is a place where children are seen being born and growing up. YouTube is providing us with a new form of personal history and family archives. A new generation will look back on this virtual playground, see themselves, and remember their past in a manner that no other generation has experienced.

6 The YouTube Wars: Politics, Religion, and Armed Conflict

War – what is it good for? Ratings. Conflict has always been good for television ratings. In the media culture of the twentieth century we participated in conflict by watching. We like to watch. Yes, we marched. We sang. We organized and protested. But most of us just watched the conflicts roll across our television screens. We are still tuned in and watching the wars du jour. As I wrote these words, Israel was once again at war. Car bombs continued to explode in Iraq, body bags were being returned from Afghanistan, India's and Pakistan's diplomats were fighting over who killed whom, and southern Africa was a deadly mess of duelling warlords. We turned on our television sets and watched, but we also started to do something quite different. We turned on our video cameras and engaged in various skirmishes online.

Amateur online video is changing the way the audience participates in conflict. In this new mediascape we act as producers of conflict. This chapter presents three case studies showing how individuals use YouTube to participate in and represent political, religious, and armed conflict. In 1997 Steve Whittaker, Ellen Issacs, and Vicki O'Day offered a description of the core attributes of virtual community that included shared goals, repeated participation, shared resources, reciprocal support, and shared social conventions.[1] Missing from this and many such definitions of online community is that most universal of human realities – conflict. There is plenty of conflict within YouTube's communities.

The First YouTube Elections

Few things are more contentious than the struggle over the emperor's crown. The American mid-term congressional election of 2006 has been

called the first YouTube election. Herein we will consider the period between 2006 and 2008 as YouTube's first election, as it culminated in the
American presidential race of November 2008. During the 2006 mid-
term election 11 per cent of American Internet users posted their own
political commentary, created a political video, or circulated someone
else's political video.[2] In other words, millions of Americans produced
online political content every day in 2006 and continue to do so. The
level of political participation in cyberspace has been observed to be
much higher. The Pew Internet & American Life Project's 2008 survey
concluded that 46 per cent of Americans use the Internet to participate
in the political process and 35 per cent of adult Americans have watched
an online political video.[3] Barack Obama's supporters were the most active users of the Internet. The overall percentage of the population that
is using the Internet to engage in political action increases with each
passing year. Nonetheless, some media scholars continue to question the
significance of online political activity.[4]

The Internet plays an increasingly large role in shaping political values and actions, because many users believe that it offers more in depth
information, perspectives, and participation possibilities.[5] The Internet
is used and perceived differently from mainstream sources of political
information. The main difference is that an increasing number of wired
citizens participate in elections by producing political content. This
greater participation, in turn, is eroding the control that campaigns have
had over their images and messages.[6]

The *Los Angeles Times*'s lead editorial of 6 September 2006, 'The First
YouTube Election,' noted that YouTube is a 'tempting launch pad for
political mischief because it is effectively unregulated by the Federal
Election Commission.' YouTube was being used by professional lobby-
ists and public relations companies that attempted to pose as amateur
videographers and by amateurs who recorded clips at political rallies and
uploaded forgotten videos from past campaigns. These clips would often
'aim to show candidates being hypocritical, ill-informed or unappealing.
The Internet can serve as an important memory bank for gaffes and public lapses that candidates and officials would rather citizens forget.'[7] One
of the earliest political consequences of YouTube is that it is expanding
the memory of the citizenry.

Amateur online video changes the way we participate in democratic
processes. Amit M. Schejter and Moran Yemini argue that under the regime of corporate broadcast media the 'masses were silenced for the
sake of the public good.'[8] They propose that the Internet has the poten-

tial to re-establish freedom of expression and maximize the active participation of citizens. Online amateur video practices may prove to be an important step towards a reinvigorated culture of political participation.

Some of the more famous moments of the first YouTube election were the result of videos made by professional production companies. Uploaded pirated copies of Al Gore's *An Inconvenient Truth* and the made-for-YouTube music video *'I Got a Crush ... On Obama' By Obama Girl* attracted millions of online viewers and tons of press coverage. Internet users spent a total of 1 billion minutes viewing Obama videos.[9] Barack Obama appears to have outmatched his rivals in gaining attention on YouTube and Facebook. The Obama campaign machine was renowned for its sophisticated use of the Internet and YouTube. In January 2009 YouTube and the United States Congress launched channels for the House and Senate that enabled lawmakers to post videos, interviews, and responses to citizens' questions.[10] In Quebec Michel Rivard's video *Culture en péril* was credited with preventing the Conservative party from attaining a majority government (an overstated claim, but one that acknowledges the growing significance of YouTube videos in elections).[11] I'll leave these professional uses of YouTube to one side and focus on amateur political uses of the 'Tube.

Politicians were constantly caught off guard by the army of amateur videographers in their midst. Virginia Republican Senator George Allen was captured on film repeatedly uttering a racial slur (*Allen's Listening Tour*). Indicative of the overlap between amateur online video and commercial media, the clip was almost immediately mentioned on the front page of newspapers and on television newscasts. This video is believed to have contributed to the defeat of Allen. Montana's Senator Conrad Burns's defeat was also blamed on the influence of YouTube videos.[12] Florida Congressional Republican candidate Tramm Hudson lost his bid for election after an amateur video that recorded him making racial slurs was posted to the Internet. As a direct result of amateur videos two 'underdogs' (Jon Tester from Montana and Jim Webb from Virginia) won their 2006 state elections. It is clear that amateur political videos on YouTube change the outcome of elections. Significantly, and perhaps not surprisingly, those videos that 'feature and expose the candidate in a negative light' generate the most viewers.[13]

In the news-stand magazine *Rolling Stone* Tim Dickinson reflected on Allen's gaffe and wrote: 'barriers to video broadcast are now gone. So an opposing campaign no longer has to rely on a local news station or CNN or CSPAN to run video of a gaffe. Any dolt with a handicam now can capture the unscripted reality of a candidate and disseminate it world-wide' (note the reference, in a commercial medium, to the amateur vid-eographer as a 'dolt').[14] This and many similar events threaten to erode carefully constructed images of public officials. As Manuel Castells has observed: 'It has become customary to post either on YouTube or similar sites embarrassing clips of opponents, sometimes recording a direct hit on the targeted candidate.'[15] As a result, amateur online videos may lead to a reciprocal change in the way politicians and their campaign organi-zations manage their public appearances.

Carter Eskew, a former media consultant to Senator Joe I. Lieber-man, suggested that amateur video practices would promote 'a kind of authenticity and directness and honesty' among politicians.[16] Amateur video practices may lead to less duplicity within political campaigns, which have often seen messages change according to the type of local audience they address. Eve Fairbanks explained to readers of *The New Republic* magazine that this may not be a welcome development among the political class: 'politicians know that different audiences – say, a huge stadium full of college-age pro-choice activists versus an intimate living-room gathering of religious seniors – want, and need, to hear different messages. Getting disparate groups with disparate interests on board with your political program requires that they hear tonally different, if philosophically similar, concepts – and that they *not* have the opportu-nity to watch, obsessively and increasingly irately, your speech aimed at a different group.'[17] The traditional style of political rhetoric requires audiences to be isolated from one another. YouTube works against con-ventional campaign practices by bringing disparate and isolated audi-ences together.

New amateur video practices such as 'tracking' – following a politician around while armed with a video camera that is always on and waiting for him or her to say the wrong thing – may eliminate spontaneity and force politicians to seek out more controlled events and avoid speaking in public. They may also lead to more accountability for words spoken and promises made. In 2006 Republicans placed a television advertise-ment in Tennessee that used sexual and racial slurs to demean Harold Ford, a Black Democratic candidate for the Senate. When the local tel-evision advertisement was uploaded to YouTube and exposed to a much wider audience, it became a national embarrassment.[18] Politicians are still adjusting to the demise of the local – all audiences now are national (and international) audiences.

It is not hard for politicians and campaign managers to see which way the virtual wind is blowing. Increasingly, elections are being fought, won, and lost online, and amateur video is playing an ever larger role in political skirmishes. Elections worldwide are now playing out on YouTube. Demonstrating how resistant some politicians can be to the changes brought on by new media, the Ontario leader of the New Democratic Party, Howard Hampton, once proclaimed that 'YouTube is not the place to communicate either policy or to communicate government messages.'[19] YouTube is used as a channel for official political communication in numerous nations. Even the Queen of England (TheRoyalChannel) and the Pope (vatican) have their own YouTube channels.

One of the earliest empirical studies of the use of YouTube during an election was conducted by Tom Carlson and Kim Strandberg in 2007. Finnish candidates were uploading campaign videos by October 2006. One of these videos included a female candidate sitting in a sauna and discussing her electoral agenda (Finns consider the sauna to be their foremost national symbol). Carlson and Strandberg note that the mainstream news media play a significant role in expanding the impact of YouTube videos on public opinion.[20] YouTube was seen as playing a marginal role in this election. The 2007 Australian federal election was also described as 'the YouTube election,' but in the end its role was greatly exaggerated by the press.[21] One could say the same thing of YouTube's role in British and Canadian elections between 2006 and 2007. Early academic studies of YouTube's ability to function as a public sphere that fosters democratic processes were inconclusive.[22]

Canadian Politics and YouTube

Canadians used YouTube to throw mud at their favourite political target in 2008. Using the same *Downfall* movie clip that lampooned Hillary Clinton, the video *The Harper Dictatorship* depicts Canada's prime minister, Stephen Harper, as a defeated Hitler, who says, 'When are people going to realize that I'm the greatest leader?' Viewed 103,316 times, the video attracted eighteen video responses and 681 written comments. Some viewers feel the use of Hitler's image is inappropriate: 'this video is obscene and disrespects sacrifice many Canadians made to knock out the Nazi regime'; to which another viewer replied, 'Get a sense of humor you twit.'[23] Liberal and Conservative party supporters debate each other across a variety of issues within the written comments. The video is posted to other websites such as DemocraticUnderground.com, where it has been hotly debated.

There is a complex relationship between online videos, blog discussions of those videos, and mainstream coverage of the same videos. Blog discussions and YouTube audiences are two aspects of the same media spectacle. They influence each other. Blogs often link to YouTube channels and videos, contain copies of YouTube videos, and host discussions of the videos. There is a correlation between blog discussion of a YouTube video and mainstream media coverage of the video.[24] Philip Moscovitch told readers of one of Canada's national newspapers, the *Globe and Mail*, about *The Harper Dictatorship* video only eighteen days after it was uploaded to YouTube.[25] The video is mentioned, reproduced, or discussed on over 1,000 blogs and video sites. Amateur video practices influence mainstream media coverage and influence the type of symbols and meanings that are used within political discourse.

The eighteen video replies to *The Harper Dictatorship* included one from a man who expressed his comments in song (*Wasting Time at the Parliament!*). None of these replies generated further replies. At the beginning of 2009 political discourse by Canadian YouTubers did not exhibit extended debates carried out via video replies.

Another version of the *Downfall* parody is called *Adolph Harper loses it* (41,597 views) and critiques Harper's proposal to cut funding to the arts. Hitler/Harper is made to say, 'these documentaries about gay marriage and homelessness, they're trying to turn Canada into communists … Why is it so important to have a culture distinct from the Americans? I'd give my left nut to have a culture like the Americans.' One viewer responds, 'Amazing satire! It captures just about everything going on right now. I love it and I'm circulating it to everyone I know!'[26] Here we see the viral effect of amateur cultural production at work. This amateur political video resonates with what people are following in the commercial press and becomes one more media spectacle passed from person to person. The comments attached to this video roughly follow along the lines of a debate about government funding of the arts. Just as newspapers and television newscasts set the agenda for public debate, YouTube videos also establish themes and issues that become fodder for online debate. Amateur online video produces and reproduces political discourse.

The leader of the opposition in 2008, Stéphane Dion, has also been subject to criticism. The video *Stephane Dion: White and Nerdy* (14,543 views) plays upon Dion's white hair, pale skin, and his reputation for being a nerd. The comments attached to the video reflect debates between Liberal and Conservative Canadians over the character and qualifications of Dion. The video is a mashup of pictures and words accompanied by Weird Al Yankovic's song *White & Nerdy*. As is often the case, various copyrighted materials are combined to create a political commentary set to music. In both Liberal and Conservative hands, bits and pieces of commercial media are remixed to create videos that challenge legal notions of fair use and cultural standards of fair play.

Noah Richler told readers of the *National Post* that Canadian political videos 'of a much more basic nature affected the outcome of the last election – and its sequel, the Rise and Fall of the Liberal – NDP coalition.' He explains the effect of the 'mass composition of videos' as speeding 'the erosion of truth as decreed by the media (always a mistake), the courts, churches and other institutions that used to be considered expert ... Effectively, we have arrived at a world that storytellers know well – one of so-called "poetic truth," in which the conventional building blocks of facts or evidence that we once relied upon may well be false but the message can be argued to be true.'[27] Richler argues that facts are losing their impact and we are losing our ability to judge the truth of a situation.

Richler's analysis accounts for only half of the story. The media, the courts, churches, and other institutions that shaped the twentieth century did not do so by mastering and relaying all the facts. The institutions of modernity told their own versions of 'poetic truth' and used these 'truths' as a form of power. They created narratives that operated as discourses of disciplining power.[28] This was the golden age of propaganda, public relations, manufactured consent, and image control. Richler would have us believe that the mass composition of videos is turning Canada into a 'Post-Fact Society.' He does not have all the facts about amateur video practices. My point here is not that amateur video is more factual or that it produces a truthful representation of the situation. Amateur representational practices are too complex and contradictory to be blamed for eroding the role of truth and fact.

The *National Post*, the other Canadian national newspaper, also has its own YouTube channel (nationalpost). One wonders if Richler would include the *National Post* in this 'Post-Fact Society' that he sees YouTube heralding. Arriving at the opposite conclusion from Richler, John Hartley argues that YouTube and online amateur video practices may herald the return to an 'ancient, multi-voiced mode of narration,' an oral mode of storytelling that promotes 'credibility, richness, and critical value' and inductive reasoning.[29] Claiming that we are heading into a post-fact society because of the mass participation in video making is a gross oversimplification of the changing character of the public sphere. All discourses of power, fact, and myth are contested in this digital public square.

Castells, along with many others, has rightly suggested that, whereas the public mind was once formed through political institutions, it is now substantially shaped by mass media systems. This arena of communication, the corporate media system, has evolved into a highly contested terrain. As the Internet community usually confounds attempts to censor the flow of information, 'dominant elites are confronted by the social movements, individual autonomy projects, and insurgent politics that find a more favorable terrain in the emerging realm of mass self-communication.'[30] Mass self-communication via video is Generation YouTube's unique contribution to media culture. In a society shaped by commercial media, mass self-communication, such as blogging and amateur video, inevitably challenges power arrangements and generates conflict.

American Politics and YouTube

Nicholas J. Slabbert has suggested that America's 2008 election 'may not really be the first twenty-first century US presidential election, but rather the last twentieth-century election, in the sense that it is an election still being conducted with the perceptions of civil society inherited from a pre-Internet age.'[31] Our vision of how civil society is constituted lags behind the new media forces at work today. On YouTube we encounter a very different kind of political conversation from the one represented by mainstream media or associated with civil society. Different types of voice are gaining public attention. Political discussion within YouTube is particularly thoughtful as well as crude, obscene, and shallow.

On YouTube we encounter voices that are not normally heard in mass media's representation of civil society. Within mass media we encounter the authoritative opinions of intellectuals, elites, and officials. The opinions of ordinary people are usually reified and presented as percentages. Political scientists and professional pollsters typically reduce public opinion to data. Words are transformed into numbers and a highly abstract representation of the mass public's sentiments is thereby produced. The methodological problems of this enterprise are well known.[32] Yet much is lost when public expressions of thoughts, beliefs, and claims are reduced to percentages and data points on a chart. Gerard A. Hauser proposes an alternative model of public opinion that treats vernacular discourse, talk among ordinary citizens, as a 'significant source of evidence deserving intense scrutiny.' Hauser argues that further insight into public opinion may be gained if we take vernacular discourse seriously. It says something about the state of opinion research that Hauser must defend the worth of ordinary conversations. Public opinion theory does give token recognition to the importance of the mundane conversation of ordinary people; yet, says Hauser, 'their discussions of public opinion seldom treat discourse seriously.'[33]

Beginning with the observation that public opinion is a product of discourse, formed through the ebb and flow of countless conversations, Hauser argues that the rehabilitation of public opinion necessitates the careful interpretation of vernacular discourse. This discourse is not the primary conversation of officials, which sees 'leading voices speaking and writing from institutional seats of power.' It is the everyday conversation of ordinary people – mundane dialogues that are not necessarily noble, free of bias or ideological distortion (109,110). Sounding very much like an anthropologist, Hauser notes that ordinary conversations express widely shared core values and meanings that are nonetheless constantly subject to debate and dispute.

Although intellectuals refer to ordinary conversations as discourse, discursive practices, and rhetoric, the people involved in these conversations would simply refer to their activity as talk. Indeed, that is perhaps the best way to describe political online amateur videos – people having a public conversation. Over the past fifty years scholars in many fields became increasingly interested in public conversation. As Hauser notes, the critical analysis of public discourse has attained a central place within the western intellectual tradition (13). It is no coincidence that the perceived importance of ordinary conversation has grown parallel to the spread of the modern media systems that have helped to capture and transmit public conversation. Contemporary critical theorists attribute considerable importance to mundane public conversations; they are a significant source of our shared meanings. Through these conversations people collectively engage in debate over the proper order of society, consider possible collective futures, define group identities, and participate in the construction and reproduction of social order.

Conversation is not the sole source of social reality, but it is a substantial force in determining the course of society. 'We cannot make sense of our individual lives,' suggests Hauser, 'without understanding how deeply discourse shapes us ... society is awash with rhetorical exchanges that contribute to the continuing cultural, social, and political education of its members' (34). Modern scholars were not the first to recognize the significance of public conversation. Over 2,500 years ago the ancient Greeks viewed public deliberation as the foundation to community, civilization, and democracy.

Hauser presents a sound argument for the relocation of public opinion in ongoing dialogue of the citizenry, but unfortunately he provides no concrete method for executing a discourse-based theory of public opinion. It is equally unfortunate, and rather odd, that Hauser's *Vernacular Voices* (1999) mentions the Internet only once and does so only to dismiss its immediate relevance. Hauser mistakenly sees access to the Internet as beyond the means of the 'average individual' and thus projects the Internet's 'enormous potential to become a significant discursive arena within civil society' into the future (293). This oversight is understandable, as the rate of growth and change in the Internet community renders the late 1990s ancient history in digital time. By December 2005 the Internet community numbered 1 billion members worldwide and 72 per cent of Americans were Internet users by 2007. By 2009 the Internet had displaced radio as the second most used source of entertainment and information (television held on to first place).

We need no longer anticipate the Internet's potential as a significant discursive arena. That future has arrived. One very small study collected a limited sampling of YouTube election videos over a period of one year (mid-2007 to mid-2008) and found over 20,000 videos, which is only a small fraction of the total number of YouTube videos in this category.[34] The actual number of YouTube videos and written comments that are political in nature measures in the millions.

The future of America's civil society is quickly unfolding within the Internet, and the latest additions of YouTube and amateur video suggest we are headed into a rather uncivil virtual society. Consider the amateur animation *Barack Obama masturbating*, which portrays Obama masturbating in the Oval office to the tune of Russia's national anthem. This use of obscenity has long been a feature of political commentary and played an important role in the Enlightenment, so dismissing this and related material as juvenile overlooks the facts and forces of history.[35] Obscene political videos are also made by professionals working within media companies that produce YouTube videos such as *I Masturbated To Sarah Palin (John McCain Ad)*. Thus, we cannot blame the obscene character of YouTube on the lack of controls on amateur production. In many instances amateur videographers are merely replicating the aesthetics of commercial media.

In the twentieth century the character of America's public debate was substantially determined by the regulatory controls embedded in the commercial media. Controlled and regulated media mediated a significant portion of what constituted the public sphere and public opinion.

What remains to be seen is how political processes and public opinion change as a result of the uncensored mediation of ordinary conversation within the Internet. These conversations are chaotic in structure, but the Internet has a way of bringing order to its chaos. Although at present the organization of YouTube's collective political dialogue is at best primitive, tools are being developed that may allow for sophisticated and highly accurate identification of the ideological perspective of YouTube videos.[36] This may enhance the utility and impact of video as a vehicle for vernacular political exchanges.

Within YouTube one of the more common methods of political discourse by ordinary citizens involves simply uploading television content and adding a title or description that frames its meaning. In these instances the preferred meanings of the original content are altered by the new titling and descriptions. In a variation a mixture of television content, annotations, still images, and a recording of the amateur videographer or other YouTubers provides some form of commentary. There are many other forms of visual political commentary that range across all artistic forms. Even fridge magnets have been used to talk about politics on YouTube (*Magnetize Dialogue*).

Like Canadians, Americans are seen using Hitler as a visual metaphor for their political enemies. The video *Hail Hitler, Hail Obama* (842,102 views) combines images of Obama and Hitler along with scenes from a sermon by a New York preacher, James David Manning, in which Manning also compares Hitler to Obama. Similar videos include *Fuhrer Obama – Grant Park, Chicago – Hitler Youth* (260,076 views) and *Barack Obama Kids and Hitler Youth Sing for Their Leader* (570,388 views). Comparisons between a politician and Hitler or the devil are common within YouTube. Since these comparisons are occasionally made on television shows such as *Family Guy*, spoken within some of America's churches, and implied by the political rhetoric of some of America's politicians, we cannot blame this particular style of political discourse upon the culture of the Internet alone.

When we wade into the YouTube community's political dialogue, it is hard not to be dumbfounded by the volume of obscenity and hatred that is readily found. American politics have been described as suffering from a pattern of self-reinforcing hatred.[37] Politicians use hate-filled rhetoric, talk-show radio hosts fill the airwaves with hate, commercial cultural products such as rap music promote hatred, and the Internet has unleashed a tsunami of hatred. Here I am not claiming that American online discourse is *more* hateful than that of other nations. Also, at this time we cannot be certain that online the right wing is more hateful than the left wing.

Hate first went online in 1984 and is not going to go away.[38] Thus far the vast majority of studies on online hatred have focused on extremist groups, yet hatred and intolerance are a significant part of everyday online conversations. It would be an overstatement to say that the vernacular of cyberspace is hate, but it is nonetheless true that hatred is a dominant characteristic of the 'Tube. YouTube has been accused in court of 'being an accomplice to inciting racial hatred and discrimination.'[39] Certainly, it has a significant problem on its hands.

My God Can Beat Up Your God

A considerable number of video bloggers on YouTube engage in debates over religion. Some of the larger areas of debate are focused on evolution, abortion, atheism, Scientology, Mormonism, Christianity, and Islam. The subject matter of these religious debates also appears to be significantly shaped by American cultural concerns. The following subsections will briefly explore Scientology and atheism on YouTube.

Scientology

Contention between the Church of Scientology and various members of the Internet community has a long history.[40] Scientology officials are constantly trying to control the flow of leaked documents and videos to the Internet but have largely failed to censor their wired opponents. In March 2008 the Church's lawyers tried to have a leaked video removed from YouTube; the video was copyrighted by the Church of Scientology and features Tom Cruise extolling its virtues. Viewed over 4 million times, the video *Tom Cruise Scientology Video – (Original Uncut)* is one of the more visible manifestations of the Church's failure to suppress Internet activists.

In April 2008 two critics of Scientology, Tory Christman and Mark Bunker, had their accounts suspended for alleged copyright violations. Mark Bunker subsequently filed counter-notifications to the Digital Millennium Copyright Act (DMCA) copyright claims and had his XenuTV account reinstated. This series of events led to closer scrutiny of the DMCA and YouTube's interpretation of fair-use practices.[41] There are thousands of videos on YouTube that are critical of the Church of Scientology. Bunker himself is credited with using YouTube to guide the tactics of a group of anti-Scientologist activists known as Anonymous.[42] The Church also had its own account suspended by YouTube for copyright violations but now pays for its own channel.[43]

In the ongoing Scientology conflict 4,000 illicit DMCA takedown notices were filed against many videos and accounts, most of which were subsequently reposted and reinstated after individuals complained to YouTube. Again, this incident drew attention to problems with the DMCA: 'the guilty-until-proven innocent method of dealing with notices like this may have to be re-evaluated. While filing a false DMCA notice is a criminal offense, prosecution in these cases rarely comes about.'[44] YouTube has been accused of having a pro-Scientology bias, but this line of argument fails to explain the enormous volume of anti-Scientology speech that proliferates on the 'Tube. At best, the Church of Scientology is fighting a battle that the recording industry has already lost. Its legal tactics only incite further individual and collective action, foster hostile attitudes, and draw more attention to whatever it is trying to suppress.[45] Given the largely hostile reception that the YouTube community has extended to Scientology, Lorne L. Dawson and Jenna Hennebry are correct to suggest, at least in this instance, that the Internet has not led to 'an intrinsic change in the capacity of new religious movements to recruit new members.'[46]

Atheists Versus Creationists Versus Everyone Else

One of the more intense YouTube wars is between atheists and Christians. Both sides of this cultural war defend their positions with religious fervour and repeatedly engage each other in protracted debates. These debates make use of material from television shows and a wide variety of other sources. They also make use of parody. Edward Current has created dozens of comedic parodies of Christian beliefs, the most popular of which has garnered over 1 million views (*The Atheist Delusion*). Atheists and creationists also face off against each other on YouTube armed with DMCA takedown notices.[47] In some instances lawyers use YouTube to alert involved parties to this type of activity (*Rational Response Squad Banned!*).

Virginia Heffernan described atheists as 'the religious group that makes the most imaginative and despotic use of YouTube.'[48] One of the more curious videos in YouTube's religion wars has been made by an activist group called the Rational Response Squad. Viewed over 923,000 times, *The Blasphemy Challenge* invites YouTubers to make a video wherein they deny the existence of the Holy Spirit and thus blaspheme.[49] The video garnered over 1,470 video replies (one of the highest number that I have observed) and 75,953 written comments. The majority of these replies receive fewer than 500 views. Among all video replies any further video replies are very rare. Nonetheless, the idea of blasphemy touches upon something at the centre of the YouTube community's culture.

Various individuals take the challenge, while others use the challenge as an opportunity to announce their faith in Jesus. Some of the most viewed responses are *Brigitte's Blasphemy Challenge* (54,481 views) by blonde Brigitte Boisselier, a spokeswoman for the Raelian new religious movement, and *God. Is. Imaginary.* (462,797 views) by eighteen-year-old Jessica. Clearly, the YouTube audience is drawn towards attractive female blasphemers! What is striking about this media spectacle is the improbability of its occurring on network television or being sponsored by a major corporation. The atheists of YouTube illustrate how marginal groups are using the 'Tube to challenge conventional thinking and fight for cultural legitimacy.

The visibility of atheism and its debate within YouTube occur in the peculiar religious context of America. In 1988 George H.W. Bush, while campaigning for the presidency, was asked by journalist Robert I. Sherman, 'Surely you recognize the equal citizenship and patriotism of Americans who are atheists?' Bush replied, 'No, I don't know that atheists should be considered as citizens, nor should they be considered patriots. This is one nation under God.'[50] The vernacular discourse of atheism within YouTube is part of a significant challenge to the normative beliefs of a Christian empire.

Here I have barely touched the tip of an iceberg. YouTube's religious wars are vast, varied, and of tremendous cultural significance. They also throw into question claims about 'the typical American reluctance to criticize or even discuss the particulars of another's religion.'[51] On YouTube any religious belief is fair game for questioning, parody, and outright ridicule.

The Virtual Killing Field

'Why don't you come on back to the war, it's just beginning,' sings Canada's poet-songwriter Leonard Cohen.[52] There is a war on YouTube over the representation of war. Largely because of the Internet, western nations have lost control over how war is represented. More to the point, conquering or invading nations are now incapable of preventing the enemy from representing its side of the war within cyberspace. Even the Taliban now have their own YouTube channel (Istiqlalmedia). This is a potentially destabilizing development, as all modern wars have been fought with a combination of military action against the enemy and propaganda directed towards one's own civilian population. War-time domestic propaganda has ensured that the citizenry see the enemy as completely evil and the cause as entirely just. In one way or another, nations always have made it clear to their citizens that there is no acceptable middle ground – citizens are either with the enemy or against them. War is not presented as a debate. The following section will briefly explore how YouTube is used to represent armed struggle.

Insurgent forces around the world have become adept at using video. These videos can be two-hour features in the documentary style with very high production values. They also include short operational videos that document specific attacks; hostage videos; statement videos (the equivalent of press releases); tribute videos that commemorate the death of significant group members; and training, instructional, and recruitment videos.[53] In all cases the production values have become increasingly sophisticated. In some instances, like a Hollywood production, insurgents stage attacks and multiple cameras are used to record the event from multiple angles. The University of Arizona's AI Lab Dark Web research project has located and archived over 15,000 videos from terrorist websites, of which more than 50 per cent are related to Improvised Explosive Devices (IEDs).[54] Many amateur and professional video productions by insurgents end up on YouTube.

In one instance an Egyptian student attending the University of South Florida was sentenced to fifteen years in prison for making a YouTube video that showed how to make a bomb detonator out of a remote-controlled toy.[55] Curiously, there are hundreds of videos on YouTube that show how to make explosives. There are also thousands of videos made by Americans that show the explosion of chemical, oil, acetylene, gas, and other types of homemade bomb.

One study in 2008 concluded that 'although YouTube has made efforts to control video content, the site is still heavily used by extremists for video sharing.'[56] After complaints from Senator Joe Lieberman in 2008 YouTube changed its content policies and deleted some videos that incite violence. Since that time the visibility of insurgent videos on YouTube has decreased. Such videos do continue to appear but are more rapidly deleted. In the wake of YouTube's new policies, insurgent videos migrated from YouTube to other video sites such as LiveLeak. As

Frank Cilluffo, Homeland Security director at George Washington University, noted, 'The reality is by shutting YouTube jihad videos down, it is more or less a game of whack-a-mole; they pop up somewhere else.'[57] Insurgents use the Internet to disseminate detailed instructions on how to post videos to YouTube in an attempt to bypass the site's censorship mechanisms.[58]

While videos made by insurgents remain 'underground' on YouTube, it is easy to locate images of insurgents being killed in Iraq by occupying forces. Also, videos of IED attacks on occupying forces within Iraq are widespread and attract millions of views. In May 2007 there were approximately 2,300 videos tagged with the letters 'IED.'[59] By November 2009 this number had swelled to 9,620. YouTube's policy is to delete videos that incite violence, yet countless such videos continue to draw millions of views.

Regardless of YouTube's policies, many amateur mashups of military footage explicitly promote violence towards entire civilian populations. Consider the example of the video *Thunderstruck – US Army* (645,728 views), which shows various American military vehicles attacking targets. The video has been on YouTube since summer 2006 and is accompanied by the text, 'What we need to start doing to the whole damn Middle East.' The video prompted a video response from another YouTube member that compared the Iraq wars to the Vietnam war, thus questioning the legitimacy of the Iraq wars (*Please Explain The Difference*).

Within YouTube there is an intense debate over the war. The debate takes many forms, such as video bloggers' talking directly to the camera or simple musical videos that combine a song with a series of pictures. The YouTube community frequently uses 'found media,' appropriated songs and images, to express an opinion. The video *Peace for Iraq – Stop the war* (95,968 views) contains images set to a song with the meaning reinforced by the video's title and its description. The description attempts to deflect criticism by explaining, 'I am not Anti-American, I am very much Pro-America and Pro-Peace. I have nothing but respect for all the brave Men and Women soldiers from any country who fight for our collective freedoms, in Iraq, Afghanistan, or elsewhere. I posted this video out of empathy for the innocent Iraqi people, but I am absolutely Anti-Terrorism, Anti-Insurgents and Pro-Peace in the world.' The video prompted over 1,500 written comments that involve a heated debate over the war. This is merely one of millions of such instances of amateur videos acting as a new form of vernacular speech, a manifestation of public opinion that incites the production of more opinions. It is not uncommon for such war-related discussion videos to garner millions of views and thousands of comments.

When Israel and Hamas were fighting in Gaza in December 2008, both sides made extensive use of YouTube to influence public opinion. The Israel Defense Force created a YouTube channel (idfnadesk). Hamas responded by creating its own video distribution site (palutube.com). Following the general lines of conflict within YouTube, Hamas supporters flagged some of the Israeli videos as inappropriate and YouTube removed them.[60]

Only a few weeks into the war *Time* magazine suggested that Israel was losing the propaganda war.[61] One of the reasons] was the use of the Internet. A new media analyst for *Al Jazeera*, Riyaad Minty, noted that, whereas the Israeli army showed black-and-white videos that lacked audio, the Palestinians uploaded 'vivid videos of the chaos and destruction on the ground following Israeli air raids.' While *Al Jazeera* may be as biased as Fox news (although in a different direction), Minty's conclusion that the Palestinians' unfiltered videos from Gaza 'allowed pro-Palestinian supporters to dominate this online war' may be correct.[62] On the other hand, Amir R. Gissin, consul general of Israel–Toronto, explained to readers of the *Toronto Star* that 'numerous YouTube videos presented to the world show how Hamas fired rockets and mortars repeatedly from schools, hospitals and UN compounds.'[63] Nonetheless, it remains unclear as to who actually won in the court of public opinion, although it was obvious that Internet video was playing a much larger role in this war.

The Vietnam war was defined by Nick Út's Pulitzer prize-winning photograph of Kim Phuc as a naked child running from her napalmed village. The second Iraq war was defined by images of Americans torturing Iraqi prisoners in Abu Ghraib prison. Reflecting on the 2009 Israel/Hamas war, Andrew Lee Butters rightly suggests to *Time* readers that 'the fighters of both sides are well aware of the need to produce what they hope will be the defining picture or video clip of the war.'[64]

In 2006 *Time* called Iraq the first YouTube war.[65] Throughout the war amateur videos produced by soldiers proved to be a constant source of humiliation for the military. In 2007 the Defense Department forbade its soldiers from uploading any more material to sites such as YouTube and MySpace but reversed the prohibition in 2009. The American military created its own video-sharing site, TroopTube in 2008, yet on YouTube there remains a 'large number of video clips showing members of the US military engaged in violent, antisocial activities.'[66] It remains to be seen if armies can control their soldiers' video cameras any better than they can control the enemy's ability to communicate to the global public. The official voices of military propaganda and public diplomacy are being overwhelmed by amateur video on YouTube.[67]

Henceforth all wars will be fought on the battlefield of YouTube. Vietnam was the first living-room war. Iraq is the first bedroom (where many of Generation YouTube's computers are located) war. It is also the first pocket (where our cell phones are) war, the first laptop war, the first iPhone war, and the first Blackberry war. Since some people even have Web browsers embedded in the door of their refrigerator, it is the first refrigerator war. The new media environment is the total environment. It is everywhere we are, and anywhere we are, our wars will come to us. In the end, it is not really about YouTube. It is about the ubiquity of devices connected to the Internet. It is about billions of digital movie cameras connected to the Internet. The YouTube wars are fought on a global battlefield where no one nation, corporation, or ideology actually controls what we do, record, or see with our digital recording devices.

No military has full-spectrum control over the information environment. No military has complete control over the media practices of its own forces. The world has seen plenty of examples of the grim reality of war uploaded to the Internet by soldiers, mercenaries, insurgents, and civilians, and that reality is constantly shocking and 'awing' us. The new media environment paints a very grim picture of any war, no matter how just its cause may be. It relays an aspect of reality to us that is otherwise denied by the heavily censored mainstream media. This brings into question the visibility of war in the Internet age.

Oddly enough, some theorists claim that war is losing its audience. American audiences for imagery from the Iraq war 'are remarkably scant,' claims Susan L. Carruthers. She speaks of a 'collective aversion to inspecting the war and its consequences,' and she may well be right.[68] Atrocities are committed by both American forces and the enemies they face, yet America's media are focused on the declining value of homes and retirement funds and the inflated threat of immigrants. Indeed, the only explicit directions thus far from the state have been to be more wary of neighbours and to keep on shopping. These are strange times.

Carruthers gets to the heart of the matter when she notes that in the summer of 2006 'opinion polls found that a majority of respondents hadn't heard of Abu Ghraib.'[69] One must also keep in mind the 24 million Americans who believe that Elvis is still alive.[70] Statistics can misrepresent a national consciousness with tremendous ease. When in

2008 Carruthers speaks of a 'collective turning-off' from the war, she insufficiently accounts for what we are tuning into online and what we are recording, uploading, and commenting upon.[71] The numbers have yet to be crunched (and may not be subject to easy quantification), but perhaps the audience is deeply involved in the war – just not in the way we normally understand and measure audience behaviour. Given the vernacular of amateur online video, old media metrics will miss much of the YouTube phenomenon.

Moisés Naím, editor in chief of *Foreign Policy* magazine, compared YouTube to a media phenomenon known as the 'CNN effect.' Naím suggests that CNN and the spread of international cable news networks brought in their wake a reduction in electoral fraud, helped to contain famine, energized democratic uprisings, and played a major role in wars. While Naím may be overstating the effect of cable news (and he certainly understates the propaganda role of CNN) he rightly notes that a novel news system did change the way the world sees events. Naím may also be correct in suggesting that 'the YouTube effect will be even more intense. Although the BBC, CNN, and other international news operations employ thousands of professional journalists, they will never be as omnipresent as millions of people carrying a cell phone that can record video.' Naím also notes that because of YouTube and the mass participation in amateur videography, 'it is now harder to know what to believe. How do we know that what we see in a video clip posted by a "citizen journalist" is not a montage? How do we know, for example, that the YouTube video of terrorized American soldiers crying and praying for their lives while under fire was filmed in Iraq and not staged somewhere else to manipulate public opinion?'[72] His answer to this question is the wisdom of the crowd. In the end, the truth of the matter tends to be revealed (remember Lonelygirl15?). Yet I am not so sure it is as simple as salvation by collective fact-checking. In spite of the Internet, legions of bloggers, and the *New York Times*, millions of Americans still believe that Saddam Hussein was behind 9/11. The facts often do not make any difference in the face of entrenched beliefs. The Internet effect cannot be reduced to a better, clearer picture of reality.

Summary

YouTube is a fierce battleground over public opinion and over the legitimacy of various groups and beliefs.[73] The 2009 election in Iran provided a powerful demonstration of the influence that amateur videographers can have on politics and global opinion. It also demonstrated the failure of the state to control the production and distribution of moving images.

The YouTube community is a politically engaged and consequential community. Both organized groups and isolated individuals try to undermine their religious or political opponents through all manner of videos that 'speak' to each other and the wired global village. Meanings are communicated through video titles, descriptions, visual and audio content, and written commentary. In amateur video practices we find a new form of vernacular speech – speech through the production of original and appropriated images and words. YouTube is a new global arena where public opinion is formed and expressed.

Along with the vernacular speech of videos, America's DMCA legislation is being used as a weapon within the world's culture wars. It has been used in all three arenas – politics, religion, and armed conflict. The DMCA is a particularly pernicious weapon in online cultural battles, as an Internet service provider such as YouTube 'is not required to investigate the validity of copyright claims, and the result can be a form of on-tap censorship.'[74] The misuse of DMCA takedown notices can harm the public interest. A manufacturer of electronic voting machines, Diebold, made fraudulent use of the DMCA in an attempt to have leaked company emails removed from the Internet. Diebold lost the case and was fined $125,000 (U.S.).

YouTube's implementation of its content policies and its response to DMCA takedown requests have been at best uneven. Israelis claim a bias against their nation and anti-Scientologists claim a bias in favour of Scientology. While a close investigation of YouTube's DMCA practices remains outside the scope of this study, there is overwhelming evidence that the DMCA legislation is grossly biased in favour of media corporations. The Internet is cluttered with an ever increasing number of very well-documented accounts of how YouTube has capitulated to DMCA requests against individuals whose video practices are well within the domain of fair use. Media scholars are also finding it difficult to upload videos that analyse film and television, as their analytical videos are often deleted from YouTube. As former film critic for the *New York Times*, Matt Zoller Seitz, notes, 'There should be a way to distinguish between piracy-for-profit (or unauthorized, free redistribution) and creative, interpretive, critical or political work that happens to use copyrighted material. And there must be an alternative to unilateral takedowns.'[75]

Amateur videographers' growing ability to circumvent official voices and directly represent events and vernacular opinion represents a threat to the carefully constructed identity of religious and political authorities. Modern institutions are the product of over a century of careful image management – the fruits of enormous spending on public relations and propaganda. YouTube's army of amateurs demonstrate time and again that there is an entirely new set of challenges for those who are intent on manufacturing consent and shaping beliefs.

New media practices typically change faster than institutionalized modes of representation. It remains to be seen how well military and electoral campaigns will adapt to mass participation in video making. From the global to the local, YouTube and amateur video are now deeply embedded media practices and touch upon all areas of life.[76]

7 The Post-Television Audience

Amateur online video has changed who can see what (almost everyone) and what gets represented (almost everything). It has deepened our involvement in the universals of shared culture and heightened our awareness of the particulars of local culture and difference. It has also changed our status as audiences and consumers. The mass audience are moving from their old analogue position as consumers to their new digitized position as producers. Audiences are watching and interpreting YouTube videos not just as passive viewers but as active commentators and as producers of their own videos. The categories that once strictly divided society into producers and consumers are becoming increasingly blurred. This single social fact has significant implications for the next stage of capitalism and its media culture.

Over a quarter of a century ago during the golden age of network television Raymond Williams described the commercial media system as consisting of 'centralised transmission and privatised reception.'[1] He was right. To a considerable but not an absolute degree, that centralized and privatized media system created homogeneous representations of social reality. Where the audience stood in relation to this system of transmission and reception has long been a matter of debate. The history of our conception of the audience is itself somewhat muddled and subject to much confusion.

According to James Curran, a standard outline of the history of communications research presents the following path to the discovery of the active audience. Media studies evolved from an early pre-1950s period, when a direct line was drawn between content and effect. During this time a message sent was thought to be a message received. This paradigm is said to have overestimated the power of commercial media and overstated the passive nature of the audience. In the post-Second World War era new methods of media research debunked these notions of a passive audience and 'hypodermic-needle' media effects. Curran has convincingly argued that this often repeated storyline mistakenly describes as an innovation what was, in fact, a rediscovery of 'the independence and autonomy of media audiences.'[2] One thing is certain: over the course of the last century some schools of thought have overstated the power of the media system and others have overstated the freedom of the audience. In this regard not much has changed.

By the 1980s the pendulum of history was thought to have swung too far in favour of audience freedom. John Fiske and cultural studies are often cited as the epitome of the tendency to overstate the power of the audience. Valerie Walkerdine claims that Fiske 'invokes an American discourse of empowerment, of voicing and authentic creation' that leads to overstating the audience's ability to make their own meanings.[3] The field of cultural studies, which has been very influential in media and communication departments, often presents a narrative that proclaims the existence of a cultural democracy. In this theory of cultural democracy the interpretive power of the audience is elevated, and the nefarious effects of advertising, propaganda, and public relations are minimized. Fiske is frequently used as an example of cultural studies' uncritical populism, but he is not without insight into the audience's ability to resist the dominant (or preferred) meanings of commercial cultural products.

Fiske recognized that popular culture – the domain of television and film audiences – is a battleground, a site of struggle. His main innovation was the proposal to shift the direction of analysis from a pessimistic focus on ideological power to an optimistic focus on the audience's ability to engage in 'semiotic guerrilla warfare.'[4] Semiotic guerrilla warfare describes the challenge to the corporate control of culture that a truly active audience represents. Fiske offers the following description of this new perspective: 'Instead of concentrating on the omnipresent, insidious practices of the dominant ideology, it attempts to understand the everyday resistances and evasions that make that ideology work so hard and insistently to maintain itself and its values. This approach sees popular culture as potentially, and often actually, progressive (though not radical), and it is essentially optimistic, for it finds in the vigour and vitality of the people evidence both of the possibility of social change and the motivation to drive it.'[5] Fiske is right on at least two counts. The dominant ideology does have to work very hard to maintain itself against countless counter-discourses.[6] Popular culture is potentially progressive.

The debate over Fiske and the active audience is centred on the latter claim.

In taking the measure of the YouTube audience, the power of dominant ideologies and the influence of commercial media need to be weighed against our ability to make our own meanings and maintain identities that are not determined by the corporate media system. If we overstate the sovereign character of the audience, then dominant discourses will disappear from view and we will be left with 'merely a semiotic democracy of pluralist voices.'[7] The problem with overstating the progressive nature of popular culture is that one must also account for the persistence and continued domination of capitalism. YouTube is fast becoming the very epitome of popular culture, which suggests that YouTube and amateur online video are also becoming a focal point for everyday resistances and evasions. When we explore the online audience's role in cultural warfare, the question shifts to the relative autonomy that amateur *producers* enjoy. The ongoing debate between Fiskean cultural studies and its detractors will find further fuel in the new productive capabilities of the online audience and amateur videographers.

Communication and media research in the twentieth century firmly established the relative autonomy of the audience from the intentions and meanings of commercial media. There is also little doubt that a guerilla war is being waged with particular intensity in cyberspace.[8] Online audiences are attacking meaning, pirating digital property, and attempting to redefine prominent commercial brands such as Barbie and McDonald's and political personalities such as Barack Obama, George W. Bush, and Hillary Clinton. The metaphor of warfare between commercial media and audiences runs throughout communication theory.

The debate over the active audience reflects a cultural contradiction within the heart of capitalism. This contradiction is seen in the coexistence of mass forms of behaviour with resistance, subversion, and idiosyncratic attempts at evading the constraints of a highly disciplined society. Capitalism's contradiction arises out of the economic and structural necessity of massified behaviour and the equally important need to foster individual freedom as a source for creative innovation. The analysis of media audiences reflects this contradiction when it struggles to define the character of media power and the extent of the audience's autonomy.

Ien Ang envisioned the television audience as 'active meaning producers.'[9] Now that audiences are actual producers of *video* texts, the notion of audiences as active producers of meaning has much more relevance for analysing the YouTube generation. Indeed, as the audiences for broadcast media were active, then contemporary audiences, the wired and wireless audiences of YouTube, are hyperactive. In what sense can YouTube audiences be said to be active, and is this activity different from television's active audiences? These questions can be explored through the meaning-making process known as appropriation.

Slippery Appropriations

The power that active audiences have to make their own meanings has been called appropriative power. As audience members we are constantly altering the original, or *intended*, meanings of texts (a movie, book, film, and so forth). When singing hymns in church along with the congregation as a child, I would often replace the words in the hymnal with my own words so as to amuse myself. I created my own private meanings from songs. This was a form of appropriation. YouTube is filled with millions of acts of appropriation that are often incorrectly seen as copyright infractions by entertainment corporations.

One of the more famous moments of appropriation on YouTube occurred during the American presidential race of 2008. Various individuals took a four-minute clip from the 2004 German movie *Downfall* and turned it into vicious parodies of Hillary Clinton, John McCain, Sarah Palin, Barack Obama, the Republican party, and hundreds of other subjects. The parodies added new English-language subtitles to a climactic scene that depicts Hitler facing the inevitability of his defeat and responding by railing against his enemies in a heated tirade.

These appropriations became so popular that they merited mention in the *New York Times*, where Virginia Heffernan explained to readers that, 'in the original scene, Hitler is told that his reign of power is over; he then deafens himself to reality, eloquently savages everyone who cost him his dreams, vows revenge and finally resigns himself to private grief. The homemade spoofs plug into this transformation just about any hubristic entity that might come undone.'[10] These amateur creations rework commercial film footage and now act as a way to depict outrage at a failed campaign, product, sports team and other sources of disappointment. There are now so many of them that the *Downfall* clip has been turned into a plea for an end to further *Downfall* spoofs. This YouTube clip, titled *That Damn Downfall Clip*, depicts Hitler in his bunker passionately asking, 'How many times do we have to see that damn "Downfall" clip? With stupid subtitles of other damn things?'[11]

As Heffernan notes, these 'slippery appropriations of Hitler's image for satirical purposes can be hard to take.' Yet for many they are just plain funny. Heffernan's response to the *Downfall* appropriations provides particular insight into the role of appropriation as a strategy of active meaning-making. After having seen the spoofs on YouTube, Heffernan found it 'virtually impossible now to watch the movie with a straight face.'[12] Here we have arrived at one of the most powerful features of amateur cultural production. After viewing the amateur spoofs, Heffernan could not watch the movie *Downfall*, without seeing it through the lens of the spoofs. The appropriated clips informed and framed her experience of seeing the original movie. Online appropriations have the power to alter our experience of television shows and movies. The domain of commercial cultural production is coming under the influence of amateur cultural production.

In the analogue world of twentieth-century television our appropriations, our acts of reinterpretation, had little effect on the mass audience because, as amateurs, we lacked a distribution system such as the Internet. Heffernan's account of her experience of viewing the spoofs shows how appropriation and meaning-making in a digital, networked mediascape acquire a new degree of power.

Historically, the power to appropriate and rewrite the meanings of commercial media has never been as strong as the story-making power of mass media. As the power of the active audience was always derived from a privately owned text, such as a movie or a book, this was the power of the weak,[13] which could only reinterpret the privately owned meanings of commercial media. The YouTube audience has other, more powerful, options.

In the ongoing debate about the extent of the active audience's power Ang notes that the romanticization of the active audience serves only to reinforce the 'liberal mirage of consumer freedom and sovereignty.'[14] When overstated, the notion of an active audience hides oppressive media effects, simple realities of ownership, and complex disciplining mechanisms that are embedded in law and public policy. Curran reminds us that media are powerful ideological forces and cautions against the 'advancing tide of revisionist argument, which overstates popular influence on the media and understates the media's influence on the public.'[15] YouTube and amateur video have introduced a new type of power that is now available to the online audience – the power to affect the way distant others experience television and film – but it has not changed the structure of ownership within the media system itself. Nonetheless, is it possible that, owing to the rise of amateur online video, the weak are not quite as weak as they once were?

The online audience is active in a fundamentally different way from twentieth-century conceptions of the active audience. The focus of cultural and media studies during the last century was on the audience's interpretative activity. With a growing portion of the audience taking on the role of producer, the audience can no longer be seen as merely inhabiting the restricted domain of privatized consumption. The audience is active in a new productive way. It is deeply engaged in rewriting the meanings embedded in television and film and is recirculating those new meanings. Yet such increased activity in itself may not liberate the audience from the influence of corporate media.

The Post-Colonial Audience

Before we proclaim victory for the audience we need to consider the contradictory effects of a digital lifestyle that is deeply enmeshed in the manipulation of corporate media. The proliferation of *Downfall* spoofs on YouTube demonstrates that, even as the online audience rewrites meanings, it does so with increased engagement with commercial media products. The Internet presents the audience with further choices, yet corporate media in both altered and unaltered forms maintain a prominent position among YouTube's menu of choices.

Ang describes the television audience as suffering from a 'delirium' of unlimited choice. The audience's 'compulsion to activeness,' their ability to make ever more choices about what to watch, serves only to further embed individuals in the corporate media system. Thus, for Ang, audiences that are driven to participate in a frenzy of choice merely serve as symbols of 'the increasing colonization of the times and spaces of people's everyday lives for the purpose of media audiencehood.'[16] Yet we are entering a period of post-colonial audiencehood. Online audiences are becoming adept at appropriating the meanings of commercial media products to a degree that simply was not possible in the analogue era. Digital audiences are now producers in their own right, creating their own texts with their own preferred meanings. Activity and choice among the Internet audience are not the same thing as they once were with the television audience.

Ang's use of the metaphor of colonization highlights the dominant/ subordinate relationship that is thought to exist between television and its audiences. The same type of power relations between producers and consumers of media does not exist among the Internet audience. Some claim that the virtual realm of cyberspace obliterates old dominations and creates a radically new state of independence and freedom. In 1996 John Perry Barlow wrote a libertarian manifesto, 'A Declaration of Independence of Cyberspace,' which proclaimed the independence of the Internet community. Barlow boldly proclaims: 'Governments of the Industrial World, you weary giants of flesh and steel,' the Internet community has 'no elected government, nor are we likely to have one … I declare the global social space we are building to be naturally independent of the tyrannies you seek to impose on us.'[17] As wonderful as his vision of a totally free virtual domain is, it is quite clear that the Internet community is not radically freed from national jurisdictions, laws, and economic power.

The Internet has been colonized by millions of business websites, some of which now stand among the most valuable corporations. The virtual spaces of cyberspace are filled with advertising, corporate news services, government propaganda, television shows, films, and all manner of commercial media products. Thus, the notion that corporate media have a colonizing effect on the Internet audience should not be quickly dismissed. Nonetheless, commercial television does not occupy the same dominant position among the online audience as it does offline.

The subtle shift in power relations that has taken place online because of the rise of amateur cultural production can be articulated through the lens of post-colonial theory, wherein the meaning of being liberated from direct imperial rule is carefully qualified. 'All post-colonial societies,' according to Bill Ashcroft and Gareth Griffins, 'are still subject in one way or another to overt or subtle forms of neo-colonial domination, and independence had not solved this problem.'[18] In his euphoria over the new land of cyberspace, Barlow overlooked the fact that the nation-state was the birth mother of cyberspace. The Internet was the child of the military-academic-industrial complex and corporations played midwife to its creation and development. Cyberspace was not a new land created ex nihilo, vacant, and waiting for the libertarian flag of freedom to be planted on its virgin shores. Whereas libertarians and cyber-utopians attempt to position the Internet as the domain of unrestricted freedom (or assume an eventual end-state of total freedom), we are on more solid ground if we conceptualize the Internet as a former colony that is struggling for its freedom.

With each passing year the post-colonial online audience discovers more ways of expressing communicative freedom. In the first years of the Internet, humble low-ASCII email, LISTSERV (mailing lists), and

Usenet newsgroups were seen as a tremendous advance in communicative freedom. Then the World Wide Web burst across the media landscape and initiated a land rush in the very place where space is infinite. Blogging software turned the awkward HTML coding that websites required into simple desktop publishing, and as a result in the late 1990s there was an explosion of writing that may be unparalleled in history. First email, then Web-based writing (blogging), and now the third phase in the evolution of online communication, amateur video – these three stages in the development of online communicative freedom attest to a continual expansion of the audience's expressive capabilities.

Prior to the rise of the Internet the audience was more properly thought of as being almost fully colonized. Indeed, the notion of colonization runs throughout media and communications theory. As it moves into the realm of the Internet, the audience can be said to assume the position of the post-colonial. It is independent, but still subject to subtle forms of domination. The antecedents of the commercial media system, its ideologies, inequities, and power relations, continue to provide the context and master stories that frame the online audience's experience. Post-colonial theory arose out of a growing awareness of the unequal economic and cultural relations that existed in imperial power relations. Just as post-colonial studies are 'based in the "historical fact" of European colonialism,' the study of the online audience must begin with the economic fact of the colonialism of commercial media.[19]

Analysing the online audience requires that we always keep in mind the colonizing process of television. In a sense, the online audience can be thought of as simultaneously inhabiting two different states – that of the colonized television audience and that of the post-colonial online audience. In the example of Heffernan's experience of the *Downfall* clip, it must be emphasized that the post-colonial experience of amateur online culture (video or otherwise) bleeds into the domain of commercial media and changes the way we experience 'imperial' forms of colonizing media. The two domains are not separate but interact in complex and as yet not fully understood ways. When seen as a post-colonial practice, one might well describe amateur online video as one of the indigenous languages of the Internet. As such, amateur video has a vital role to play in the decolonization of the mass audience.

Life after Television

I have dim memories of the first time I heard people in the street, back in 1989, mentioning email. Seeing this new form of communication rapidly evolve into a ubiquitous feature of everyday life heightened my interest in novel forms of communication and media consumption. So I was quite surprised in 2008 when I heard a young industrial designer, Chantal Trudel, describe her viewing habits as 'post-TV.' She was referring to the use of DVDs, the avoidance of commercials, and all the non-traditional forms of television viewing that her generation is practising.

Speculation about the end of television or a post-television world is not new. Regardless of whether or not television is dying, that people are describing their viewing habits as 'post-TV' sheds light on the audience's new relationship with online video. In 1994 the futurist George Gilder wrote *Life After Television,* a typical piece of techno-utopianism that Geert Lovink calls a 'fairy tale.'[20] Gilder equates capitalism with democracy, extols the virtues of visiting distant dying relatives via interactive television, prophesies that computer networks would give 'every hacker the creative potential of a factory tycoon of the industrial era and the communications power of a TV magnate,' and argues that the personal computer will make people 'richer, smarter, and more productive,' transform capitalism into 'a healing force,' and usher in a life beyond television where the computer will be 'a powerful force for democracy, individuality, community, and high culture.'[21] Gilder's vision for the future of cyberspace is merely an idealized version of American culture.[22]

Gilder did outline the basic features of the Internet – enhanced communication and creativity. Similar to a line of thinking found among some critical theorists, he also saw television as 'a tool for tyrants.' As is often the case with American futurists, Gilder believed that technology would soon deliver prosperity for all, eliminate the inequities of capitalism, and get rid of the bad guy, which in this case was television. Not one to shy away from overstatement, he proclaims that television's 'overthrow is at hand.'[23]

Gilder provides a splendid example of liberal telecommunications policy discourse – a body of theory that looks towards the liberalization of telecommunications markets but ignores questions of power.[24] Within this theory new technologies simply make power relations disappear, which allows the theorist to proclaim that a brave new world is nigh. Gilder provides a cautionary example of an ideologically laden prediction of television's demise. Nonetheless, we consider the possibility of a world where television is displaced from its pedestal as the dominant storyteller and as a primary force of social organization.

Critical theorists such as Sean Cubitt look forward to the formation of 'a new dominant mode of visuality.' Cubitt suggests the possibility of a future where video supersedes television as our main source of stories and paves the way to a cultural democracy.[25] While it may be the case that at some point in the future television will not hold the same dominant position within screen culture as it now does, it is far too early to make any grand predictions about its imminent demise. Contradictory and unknown effects will emerge from the rise of online video that will render long-range forecasts uncertain at best.

The introduction of a new media form can actually foster growth in an older form of media. Both television and video were once seen as dire threats to film, yet television increased the number of films available for consumption and video 'encouraged a renaissance in cinemagoing and cinema building.'[26] It is abundantly clear that the relation of film, television, video, and the Internet to each other is rapidly changing. Audiences are constantly shifting to new viewing platforms such as laptop computers and even cellular phones, and content from each medium is bleeding into the other. Perhaps no word better characterizes the new position of the audience, both online and off, than *fragmentation*.

Any teacher faced with discussing television or film with an audience of high school or university students is keenly aware of just how that audience is fragmented. As a baby boomer, I have no trouble discussing television shows from the 1960s and 1970s with other boomers. For the most part my generation watched many (but certainly not all) of the same television dramas and situation comedies. What shows we did not watch we probably knew enough about to discuss or at least appreciate their significance.

Conversations with my peers about television stand in stark contrast to my conversations about contemporary television with university students. With their attention scattered among video games, Facebook, YouTube, the rest of the Internet, and a '500 channel' universe of national and international television programming, university students share so little common ground in their viewing habits that attempts to use television shows to make a point often turn into a frustrating exercise in futility. What was once a highly centralized mass audience is now an array of smaller viewing collectives, as increasing numbers of cable channels spread the audience across multiple content options.[27]

Peter Sealey's personal reflections as a Procter & Gamble marketer in 1965 indicate just how far we have moved from mass to segmented viewing habits. Sealey relates that, when working on the Crisco shortening advertising account, he could purchase three sixty-second black and white commercials on soap operas and reach 80 per cent of the women in America. He calculated that it would take ninety-seven prime-time commercials to get that same reach today. In the early days of television families tended to view the same programs together and only occasionally was the channel changed from one network to another. Many marketers and advertisers fondly remember this time of easy audience reach. Today, such highly unified mass audiences are rare. Presently, the largest networks enjoy only a 6 to 10 per cent share of the American television audience, compared with 25 to 27 per cent forty years ago.[28]

There are those who deny that the audience exists in a fragmented state. Such denials do not explain why the issue of audience fragmentation stands so prominently within the discourse of executives and business leaders of the advertising and television industries. Consider the comments of Guylaine Saucier, a former chair of the board of directors of the Canadian Broadcasting Corporation: 'A multi-channel universe has led to fragmented television audiences. In turn, that means that all conventional broadcasters, from CBC to CTV to NBC, are losing audience share to specialty channels. On the new media front, innovative technologies such as the Internet have opened a communications floodgate unimpeded by the regulations that govern radio and television. The danger in these developments is not so much that our voices will go unheard – but, worse, that they will be lost.'[29] National broadcasters such as the CBC fear that their projection of national identity will be lost to an audience set adrift in a highly fragmented multi-channel universe.

In 1990 *Time* magazine published an article titled 'Goodbye to the Mass Audience,' in which readers were told that 'the era of the mass TV audience may be ending.'[30] The reasons given for this end tell us much about the attraction of amateur online video. Programmers and networks were thought to have lost touch with viewers. Increased competition, boredom with the sameness of programs, and that great frustrater of advertising executives, the remote control, all were seen as reasons why the networks were losing audience share. The growth of amateur video could further fragment and erode the overall audience share of television.

Amateur video may be closer to the audience and better at representing emerging tastes simply because it is made by the audience. It certainly creates increased competition for eyeballs, provides an alternative to network and cable programming, and resists the tendency of studios to settle on formulaic productions and sameness. Amateur video's proliferation of quick thrills and brief clips also feeds into an attention-deficit generation. So there are plenty of reasons why amateur video will be a source of frustration to television and advertising executives.

Fragmentation in itself does not spell doom for television. In the end, even though the contemporary audience is highly fragmented, it is still watching commercially produced entertainment. What is as yet unknown is whether or not the Internet is causing individuals to watch less advertising-supported content and commercially produced entertainment. According to one study by Jeffrey Cole, head of the UCLA Internet Project, Internet users watched five hours less television per week in 2003.[31] This trend held true in every country surveyed. Year after year Internet users consistently report that they spend less time watching television since they began using the Internet. Unfortunately, what these studies have not determined is the amount of time spent online viewing television and film. While Internet users are spending more time online and less time in front of the television, we do not yet know if this has led to a net decrease in the overall amount of time spent consuming commercial media products.

YouTube provides an example of the complex relationship the online audience has with television. Consider the following listing of a day's ten most viewed videos in the fall of 2008: an advertisement for World of Warcraft, an amateur commentator called Philip DeFranco, a television clip of Sarah Palin, another advertisement for Warcraft, a CNN news clip of President Bush, Philip DeFranco again, a comedy show clip, a soccer match clip, a sixteen-second advertisement for a Britney Spears album, and a clip from *America's Next Top Model*. The YouTube audience is most certainly engaged with advertisements and television shows. It is noteworthy that the two clips that feature politicians are very unflattering. It is likewise noteworthy that the humorous commentary on the day's news by a twenty-two-year-old amateur video blogger, Philip DeFranco, is found twice among the day's top ten.

The media industry is still trying to figure out the best way to get the online audience to pay attention to advertisements. They do not always avoid advertisements and sometimes actively seek them out, yet this level of response is not enough for the commercial sector, which seeks a much higher level of engagement with advertising among the online audience. One thing is certain: the advertising and content industries are facing difficult times.

The television industry may be blindsided by the Internet in much the same way that the music industry was caught off guard by digital piracy. According to a 2008 six-country survey by IBM, 36 per cent of the online video audience watch 'significantly less' television.[32] Yet the evidence surrounding current television viewing trends is contradictory. Studies by the Nielsen company conclude that Americans are watching more television and that television viewing is at an all-time high.[33] This much is certain: television and movie viewing is dramatically shifting to the Internet.

In 2005 Blockbuster chairman and CEO John Antioco told readers of the *Wall Street Journal* that the video rental industry 'is in the tank.'[34] In 2009 Martin Peers told *Journal* readers that the recession 'may be accelerating a structural change toward free or low-cost Web video.'[35] At stake for movie companies is the 75 per cent of their $19 billion (U.S.) annual feature-film revenue that comes from post-theatre sources – DVDs and television – the very stuff that Internet users love to pirate. Meanwhile, on the same day as the Peers article appeared, a *New York Times* article proclaimed, 'Digital Pirates Winning Battle with Studios.' Both papers suggested that the movie industry might soon be 'Napsterized' – overcome by digital piracy. In 2008 'DVD shipments dropped to their lowest levels in five years.'[36] Another survey concluded that the Web 'is a growing destination for watching television,' reporting a 50 per cent increase in America's online television audiences between 2006 and 2007.[37] Meanwhile, the comments of media executives provide a stark contrast to the changing habits of their customers. Barry Diller confidently declared in 2009 that 'movies will be much less affected' by the Internet.[38] That is a stock market bet I would not be willing to take.

Anecdotal stories about individuals like David Title, who 'has dropped his cable subscription and moved to watching most of his television online – free,' appear in the news and from my own students with increasing frequency.[39] The inauguration of Barack Obama provided a taste of things to come. It was the first time a high-profile event was watched live by more people on the Internet than on television.[40] In the late 1990s it was assumed that Microsoft would dominate the transition of the Internet to television. It utterly failed to do so. For now it looks like television is migrating to the Internet. The overall picture is less than clear, as the television industry is re-engineering a new generation of television sets that connect directly to computers and the Internet in a way that the masses will readily adopt.[41] Meanwhile, content and audiences are scattering in all directions.

All of these early smoke signals need to be interpreted in light of America's habit of watching 200 billion person-hours of television each year.[42] This will be a hard habit to break. YouTube will not necessarily make it any easier, as its business model is based on delivering more, not less, television and commercial entertainment to its audiences. While appearing before a congressional subcommittee, Chad Hurley, CEO of YouTube, denied that the company was intent on competing with television. He pointed out that CBS had publicly stated that YouTube had 'helped increase their ratings by 5 percent.'[43] Regardless of Hurley's claims, which have been regarded with suspicion, YouTube may be evolving into television 2.0.[44]

Thus far it appears that the type of post-television world into which we are moving is one where the family gathered around a living room set is disappearing and the mass audience is fragmenting into smaller audiences of specialty channels and niche programming. Where television is viewed is relocating from fixed, home-based devices specifically dedicated to television viewing to multipurpose platforms such as home and office desktop computers, laptops, cellular phones, gaming devices such as Xbox, and a wide variety of portable consumer electronics. This dispersion of television throughout the geography of daily living is giving individuals a sense that they are living a 'post-TV' existence. A growing number feel that television does not play the same role in their lives as it once did. Since YouTube users are creating more content every six months than sixty years of continuous broadcasting by the three American networks have done, there is every reason for the audience to feel that a post-television age is on the horizon.[45]

If the online audience turns away from television, then television will lose its position as the primary arbitrator of tastes. It certainly is no longer the sole screen in households. Television and cinema also are no longer the only sources of moving images; they have failed to create anything other than the weakest form of a semiotic democracy – a place where people's reinterpretative powers were relatively weak in the face of the intended meanings of the producers. It is quite possible that the Internet is laying the foundation for a much stronger form of cultural democracy.

As long as the twentieth-century audience was involved with reinterpreting television, they continued to inhabit the position of the weak. But something new happens when cultural products are transformed into digital objects and when the audience is transformed into producers. The mass dissemination of audience-produced cultural goods drastically alters the audience's meaning-making capabilities. On the one hand, the audience members have gained an increased degree of transgressive potential. We can rip, mix, and burn dominant culture and reconfigure it to our own tastes. On the other hand, we can do our own thing – be our own storytellers and make our unique amateur online videos. Thus, in their new position, online audience members have two hands that work against both the designs and the intentions of commercial media.

The Essence of Video

The difference between television and YouTube is obvious and indisputable. Within the domain of television the audience is relegated to viewer and given the weak power of interpreter. Within the virtual domain of amateur video the audience is transformed into creators of words, images, and meanings. In video theory the difference between video and television is less clear. These two categories share too much to be said to exist in a simple state of binary opposition (and besides, how many times do we have to rediscover the fallacy of overly simplistic binary oppositions?).

Theorists tend to see television and video as either two distinct practices or as interdependent technologies. As James M. Moran notes, they 'share so many likenesses that we recognize them as sibling media.'[46] Video is also related to the mediums of film, photography, digital cameras, computers, and even cellular phones. What is actually specific to various forms of representational media has been the source of a century-long debate among scholars. As we move further into an era of mass participation in amateur online video, the question of what is unique to video in general and amateur video in particular will be a matter of much debate. As Cubitt observes, the 'notion of medium specificity is a central one in understanding the development of video as a cultural form.'[47] We need to know what is specific to video if we are to understand its impact on the social order. But video is not necessarily a discrete practice – it depends upon other technologies and is interwoven into other modes of representation such as film and television. This interdependence makes it difficult to discern where video practices end and television begins, and it also makes the power relations within and between these two media complex.

In the early 1990s Cubitt saw video as a new cultural form that was a victim of 'an orchestrated attempt to close down all avenues of experiment and innovation that are not consistent with key factors of the dominant constellation of ideologies.' Pre-Internet video practices were standardized as a result of economic forces and cultural policies 'dominated by an outmoded managerialist notion of efficiency' (136). These forces of domination and standardization are not totalizing. Cubitt recognizes the importance of video as counter-documentation, an alternative to official records of events, and he sees great hope in video as the foundation for a new form of cultural democracy. Cubitt rightly asserts that, in the end, there is space for audience power in video, in spite of attempts to bring it into the ideological fold of capitalism.

Video can be thought of as a new form of TV power – it gives the audience a type of power that was formerly the exclusive domain of the commercial media sector and its moguls. As we move further towards a post-television media culture, the nature of this new power reveals itself with greater clarity. Cubitt provides an interesting insight into the way that amateur videos, such as the recordings of the Rodney King beatings in Los Angeles and the Zapruder film of the Kennedy assassination, are paradoxically more credible than television's reproduction of reality, because of the hand-held tremble, the lack of edits, and the difficulty of discerning the action. 'Clarity and certainty may be a criterion of truth,' notes Cubitt, but after 'immersion in video culture' the audience fails to believe in television (150). Perhaps increased exposure to the new modes of representation found within amateur video will cause the audience to question the slick, well-packaged presentation of truth that is delivered by corporate media.

Cubitt's comments led me to reflect on the first time I saw the now infamous video diaries of Lonelygirl15 on YouTube. This popular video diary presented a twenty-two-year-old professional actress posing as a home-schooled sixteen-year-old who was making her own videos and posting them to YouTube. I was initially impressed by and then highly suspicious of the slick editing and even the wash of light that illuminated the background of Lonelygirl15's bedroom (which was actually her producer's bedroom – another deception). These production values were, in fact, the creation of a professional video production company. It did not take long before the YouTube audience detected and denounced the fraud.

We have seen the development of genres of movie that attempt to mimic the realism of amateur videos. The films *Blair Witch Project* and *Cloverfield* used a lack of production values to arrive at a feeling of amateur video's realism. Styles of representation are shared by both television and video, and increasingly so. Thus, we cannot simply declare that amateur video has a style, a set of practices, or an aesthetic, that better represents truth and reality than television does. Indeed, within YouTube it is becoming increasingly difficult to distinguish amateur from professional productions. The video industry is getting better at mimicking the amateur mode, while continual improvements in domestic video technology make it easier for amateurs to emulate professional production values. YouTube, television, and film all provide numerous examples of how difficult it can be to distinguish between amateur and commercial modes of video production.

A post-television era that is shaped by the pull of amateur video, the push of fragmented audiences, and the multiplication of representational technologies brings in its wake new everyday practices and power relations. They are poorly understood and suffer from overgeneralizations when they are seen as the result of simple binary distinctions between television and video or between amateur and professional video. As Moran notes, a 'residual aesthetic of negation' infects much of the scholarly analysis of video. Video is too often seen in terms of what television and film are not, too often seen as technologically, aesthetically, and ideologically at odds with commercial television. It is either framed as a 'regressive medium of consumption and transmission' or a 'progressive medium of production and transformation.'[48] The multiple uses of video by amateurs demonstrate that video can be a medium of consumption, transmission, production, and transformation without necessarily privileging or essentializing any one of these outcomes.

The specificity of video – the qualities that are unique to this mode of representation – will have a significant impact upon our lives. As it is closest to our lives and often undergoes the least amount of manipulation, it is tempting to think that the mode of amateur video is best defined as transparent and reality based. This line of argument has been called essentialist and technological determinism; it assumes that a technology has inherent properties and outcomes that are determined by the technology itself.[49] Moran suggests that our impression of the transparent or reality-based character of video is not a property inherent to the medium but is 'an effect of human psychology and aesthetic ideology.'[50] We are perhaps better off thinking of the specificity of amateur video not in terms of its closer correspondence to reality but in terms of its disentanglement from the constraints of the commercial and professional modes of production. As a representational tool, what most defines the specific quality of amateur video is that it is in amateur hands. Amateurs play a very special role in capitalist social orders that are structured by the dominant forces of professional, commercial modes of cultural production. Amateurs provide an alternative.

The Alternative

As it originates outside the advertising-driven industry of content production, amateur video offers something for the audience that television does not, and perhaps cannot, provide. We look to YouTube to provide us with an alternative experience to that which is delivered by television.

Even so, this claim must be carefully qualified, as YouTube is filled with millions of clips from television and movies. Like many others, I have made use of YouTube to watch clips of *Sex in the City, Family Guy, The Daily Show, Friends, Robot Chicken,* and other popular television shows. Hulu, another video website, does not accept amateur video and specializes in hosting content only from television. Delivering 437 million videos to Americans in September 2009, Hulu demonstrates that there is a demand for television content on the Internet. Yet in the same month YouTube delivered 6.6 billion videos, which suggests that the majority of the online audience seek an unfiltered mix of both amateur and professional content.

We live in a world that is substantially defined by commercial cultural products such as advertising, movies, newspapers, magazines, television shows, books, and video games. The almost overwhelming presence of commercial cultural products in our lives gives mass participation in online amateur video special significance. In the case of videos made by amateurs, we often find ourselves in a very different world from the one constructed by the commercial culture industry.

My use of the phrase 'commercial culture industry' should not be confused with the rather antiquated notion of the 'culture industry' as proposed by Theodor Adorno.[51] Adorno, along with Max Horkheimer, grossly overstated the definitional power of commercial culture when they described it as an 'iron system' that ruled with 'absolute power.'[52] Although some theorists tend to dismiss Adorno and Horkheimer (along with the Frankfurt School's body of conceptual tools), both continue to stimulate thoughtful research into contemporary culture. As Shane Gunster rightly points out, 'The culture industry has never been more powerful, never been more invasive, and certainly never been more dominated by the commodity form than it is today.'[53]

Herein the phrase 'culture industry' is used simply to indicate the production of cultural products by the institutions that constitute the content and entertainment sector. We normally see this sector referred to as commercial media or the entertainment industry. Although the business and popular press seldom attach the word *culture* to these sectors of the economy, media theory almost uniformly recognizes that what this sector produces is shared meanings that lead to mass patterns of consumption.

In contrast to commercial media, the domain of amateur video is a world defined by the alternative. It is in many ways a world of media production and consumption that is alternative to traditional forms of television viewing. Often, but not always, it is alternative in its origins, intent, economy, meanings, subject matter, style, and technique. Amateur videos are produced outside the institutional structures of television and cinema and often explore styles of storytelling different from the expected ways of being on television. Thus, the analysis of amateur video will benefit from drawing upon the field of alternative media studies.

The study of alternative media usually focuses on media projects that are explicitly political, transgressive, or oppositional in their intent. This intentional political activity is usually done by self-described activists within formal or informal institutional settings, such as those seen in the domain of IndyMedia (independent media centres). My purpose herein is not to exclude this type of intentional, activist media activity from the domain of alternative amateur video. The notion of alternative and political media activity can be used inclusively so as to also recognize the political nature of all uncontrolled representation in a highly controlled media culture. No matter how mundane the subject matter or apparently apolitical the intent, the very act of being disengaged from commercial media has political significance. In the midst of the feminist revolution of the 1970s the domain of the personal was incorporated into the political. Likewise, the analysis of amateur video needs to recognize that the mundane and the banal also belong to the political.

Chris Atton observes that the domain of the 'alternative Internet' has led to 'new ways of thinking about what it means to be a creator.'[54] The Internet and related digital technologies have certainly led to more ways of being creative. More people are creating words and images now that the productive capabilities of digital technology are so widely dispersed. Amateur video production may also bring in its wake new ways of thinking about the political implications of mass creativity that have not been wholly incorporated into market institutions. In amateur video we see the deinstitutionalization of cultural production. A large volume of entertainment now reaches the audience from outside market-based institutions. Acts of deinstitutionalized cultural production such as amateur video often produce alternative meanings – counter-hegemonic discourses. Whenever we step outside the domain of commercial cultural production, we further distance ourselves from its ideological effects.

Corporate content does the political work of capitalism – such content is essentially propaganda on behalf of the economic system. Content that originates from the commercial entertainment industry helps to sustain and reproduce consumer culture. As a form of alternative cultural production, amateur video derives its political character from its relative absence of commercial meanings. Likewise, in the context of a pro-capitalist media system, amateur video is political in that it represents an alternative use of time. To a certain extent, time spent watching YouTube is time spent not watching television.

Within the corporate media system the only legitimate audience activity is that which involves viewing commercials. Commercials do exist in and around YouTube videos in the form of banner advertisements and via an increasing amount of sponsored videos. Nonetheless, from the perspective of media corporations the single biggest issue surrounding user-generated content and amateur video is the need to 'monetize' this free content through the injection of advertising. Thus, another way of conceptualizing the domain of the 'alternative' is found in the demand from media corporations that the audience continually be subject to advertising-based content. Amateur video is alternative because it stands partially outside the dominant advertising-content relationship.

Perhaps the main feature that distinguishes amateur video as a form of alternative media practice is simply the fact that it is done by amateurs. The categories of amateur, professional, non-commercial, and commercial are not absolutely mutually exclusive – hybrid forms of online video production are found that combine all four qualities. There are complex, contradictory, and fascinating interrelations between amateur video and the professional world of television and movie production. Barbara Klinger's exploration of film, home cinema, and amateur online filmmaking leaves little doubt that amid all the disruptive forces of new media practices there remain many continuities between the amateur and professional modes of production.[55] Yet differences between these two domains do exist, and they are readily recognized by the mass audience as two separate domains (indeed, it has become somewhat of a game within YouTube to ferret out professionally made videos that pose as amateur productions).

What must be recognized is that there has been a vast increase in the volume of amateur video production, a vast growth in the number of amateur videographers, a vast increase in the size of alternative media's audience, and a significant increase in the technical capabilities of amateur video. These coordinates of 'vast' and 'significant' are not easily quantified, but they are historically unprecedented. While alternative cinema and home movies have been with us for almost a century, it is not unreasonable to assert that in YouTube and related phenomena we see a radical shift in the ways and means of creating video. The YouTube phenomenon is part of a larger social shift, which could well be labelled the democratization of the lens. Hundreds of millions of people are now in the once rare and privileged position of making moving pictures.

More is at stake in the rise of alternative amateur video production than a simple increase in volume and participation. The practices of amateur videomaking represent ways of visual reproduction and storytelling that are both similar to *and* different from those of the mainstream production studios. To borrow the words of Atton on alternative practices of journalism: within amateur video we find a form of alternative media that 'may be understood as a radical challenge to the professionalized and institutionalized practices of mainstream media.' Even when the videographer or YouTuber is simply filming the joyous laughter of an infant, there remains a curious relationship between amateur video and radical online journalism. Atton notes that IndyMedia journalism (radical online journalism) attempts to place the power of telling society's stories 'into the hands of those who are more intimately involved in those stories.'[56] This is precisely what happens within the domain of amateur video, whether its intent is radical or not.

As Laurie Oulliette suggested in 1995, 'The notion that ordinary people might make TV instead of just watching poses a potentially subversive threat to capitalist control of the media.' Back in 1995 Oulliette conceded that 'activist media has always been a marginalized activity, confined to small groups of practitioners and small audiences. Similarly, incursions of everyday images into mainstream media have been limited to a few special occasions.'[57] How things have changed. The power of telling stories via moving images, a power that up until now has been the almost sole domain of professionals and corporations, is in our own hands. Our own hands, our own lives, our own stories – now *that* is radical, alternative, and potentially politically and economically disruptive.

The study of alternative media usually focuses on the modes of media production that are found within new social movements. Within media and cultural studies the 'alternative' is most often conceived of as the domain of activists, their social groups, and the nascent institutions that collectively comprise a new social movement. Atton rightly insists that alternative media practices are not to be understood 'solely in relation to political activism.'[58] Given the expectation of corporations that the audience be continually engaged with advertising-supported content and given that the corporate media system itself has been described as anti-democratic, then there are grounds for repositioning the notion of alternative media so that it includes the production of mundane everyday reality without any explicit political intent.[59]

Such an expansion of the notion of alternative media activity may offend some scholars who, as Atton notes, desire secure boundaries around their field of study.[60] Others, such as Clemenca Rodriguez, might prefer the term 'citizens' media.'[61] Rodriguez claims that alternative media have failed to gain any significant distance from the dominant media culture. In response to Rodriguez, Atton asks if any media project can really be fully 'independent of abiding structures, whether manifested in the political order or in the micro-organisation of everyday life.'[62] Thus, we cannot ignore the possibility that amateur video is an alternative mode of cultural production that is deeply entwined in the political order of capitalism and the dominant ideologies of the day.

Summary

Active audiences are not what they used to be. With the shift from interpretation to production, the audience gains new types of representational and appropriative powers. Amateur online video can change the way we experience commercial media, remember the past, and think about the future.

Online amateur video cultivates a post-colonial audience. The Internet's post-colonial audiences have acquired a new form of partial autonomy from commercial media. Commercial media continually attempt to reincorporate the online audience into an imperial context of highly controlled content and meanings but may be fighting a losing battle. The audience has left the building. As a result, an ever increasing number of people are experiencing a sense of being in a post-television world.

A post-television period of culture does not necessitate that television disappear from our living rooms and our lives. A post-television era may simply be brought about by the multiplication of alternatives and the fragmentation of audiences to the point where there are few shared stories to discuss or dominant viewing devices around which to gather. The current discussion of cable-cutting (cancelling a cable television subscription) within the industry and the increasing use of the prefix *post* to describe our relationship with the boob tube tells us that the identity of television is quickly changing.[63] Our television viewing habits are being shaped by online video, amateur videographers, and forces that are largely beyond the control of studio executives.

Describing amateur video as an alternative practice should not be taken to mean that it necessarily provides a superior representation of the truth or reality of a situation. Moran is particularly enlightening on how the transparent transmission of reality is an inadequate way of defining the differences between television and video: 'The important issue should not be to define once and for all which medium is better suited to the task of transmitting or transforming "raw" reality but to trace how each medium signifies and constructs notions of "reality" itself according to adopted cultural codes.'[64] Regardless of the epistemological complexities that amateur video presents, it is hopeful that the last millennium came to a close as the global distribution of words was being democratized and this new millennium began with the democratization of moviemaking. The production and representation of reality is now in the hands of amateurs. Reality may never quite be the same again.

Conclusion

A popular quote attributed to Plato is circulating around the Internet these days, 'Those who tell stories also rule society.'[1] It would be a bit of a stretch to claim that amateur videographers are about to displace the centres of power and their storytelling machines. It would be equally extraordinary if mass participation in videography did not disturb and disrupt the highly controlled representation of people, things, and events. Ordinary people are making extraordinary videos and they have captured our attention. Also, extraordinary people are making very ordinary videos of mundane reality and everyday life and shattering cherished myths about ourselves.

The Internet and its newest darling, YouTube, are changing the audience's relationship to commercial media. Mass involvement in videomaking could deepen our relationship with commercial media and enhance its main economic function – the socialization of consumption practices. It could also tear us away from the centres of corporate storytelling and capitalistic persuasion. All across the media industry the story is the same. The audience is slowly drifting away from traditional sources of information and entertainment. As sources of information, magazines, newspapers, and local television news (with the exception of cable network news) all are in decline. One media analyst called this the 'flight from top-down authority.'[2] Individuals and online social networks are supplementing people's use of traditional media and threatening their dominant position as tastemakers in consumer society.

The mass production of amateur video joins an ongoing surge in user-generated content. This new age of amateur cultural production is deeply entwined with commercial media. We imitate the aesthetics of the professional entertainment industry. We aspire to join their ranks. We steal and appropriate commercially produced content. So we have not entirely fled from the television set – we are watching it and playing with it on our computers. Lev Manovich raises a central question regarding YouTube's long-term effect. Will the increasing engagement with commercial media that accompanies amateur video practices 'mean that people's identities and imagination are now more firmly colonized by commercial media than in the twentieth century?'[3] The short answer to this question is that it is simply too early to tell.

For theorists of the Marxist school, YouTube and our productive activities represent free labour.[4] It is an argument that has some merit – we are doing the work of capitalism when we help to spread the meanings and images of commercial content through our remixes and appropriative activity. Yet clearly this is not the type of free labour that most of the media industry are willing to embrace. When amateurs labour in the fields of cyberspace, they do not merely reproduce the preferred meanings of the marketplace. They twist meanings, bend copyrighted content, fold and mutilate brand images, and frequently produce a counter-discourse that privileges their own preferred meanings. It remains to be seen what effect this has on capitalism's destructive mode of production and our own destructive mode of consumption. Our 'free labour' in the media factories of cyberspace may come with a surprising consequence: the transformation of capitalism into a better way of creating and consuming.

Much amateur cultural production takes place outside the economic system. Some amateur videos are used by YouTube to sell advertising, and yes, YouTube does gain a 'worldwide, non-exclusive, royalty-free, sublicenseable and transferable licence to use, reproduce, distribute, prepare derivative works' of your videos – which means they can use your uploaded amateur videos for profit and not return a penny of it to you.[5] Nonetheless, YouTube users do retain all their ownership rights. Henry Jenkins suggests that the gift economy of amateur digital production 'could be seen as a way of economic exploitation as they outsource media production from highly paid and specialized creative workers to their amateur unpaid counterparts.'[6] Perhaps so, but when I enjoy the free music of a songbird in the forest, I am hardly exploiting the bird. When I create playful videos for my friends or for my own pleasure on YouTube, I am not suffering in the grip of an unequal relationship. If we confuse play and free expression with real exploitation, we do a disservice to those who truly suffer. The creators of *An Engineer's Guide to Cats* are not exploiting Hollywood workers or cats.

Amateur cultural production is co-opted by the marketplace but is beyond the complete control of corporations, laws of copyright, and principles of private property. Amateur cultural production does not have to submit to the imperatives of economic exchange. There is no state or corporate mechanism that can ensure that amateur cultural production will serve as propaganda for democracy or capitalism. As a meaning-production system, amateur online video is free from economic and ideological control mechanisms. It is free to produce difference.

The difference that amateur video produces has been described by Mirko Tobias Schäfer as 'bastard culture' (which is curious, as video itself has been called a bastard child of film and television). 'Bastard culture' is a term that attempts to frame participatory culture as an illegitimate offspring of commercial cultural industries. Schäfer suggests that YouTube represents 'the rise of powerful corporations shaping and controlling cultural production and its preconditions.'[7] To a certain extent this is true, but if we overstate the degree to which YouTube shapes and controls our ability to make amateur online videos, we lose sight of the simple fact that no one corporation exercises substantial influence on the overall global production of amateur online video. Ever since Adorno and Horkheimer denounced amateurs as irrelevant and meaningless, media theorists have had great difficulty reincorporating amateur culture into the centre of history. Powerful corporations are proving to be very ineffective at controlling meaning, digital property, or cultural production within the Internet. All centres are being displaced.

By placing the emphasis on amateurs as the bastards, the metaphor does not do justice to the complex history of commercial cultural industries. Many areas of the entertainment business and media sector benefited from capitalizing on earlier non-market cultural activity.[8] From the perspective of history, the bastard – and the original thief – is the marketplace itself.

The hybrid economy of amateur cultural production is deeply troubling to what Lawrence Lessig has called the 'read only' culture of commercial media.[9] Read only culture resists the advances of non-proprietary culture at every turn. The content industry resists the audience's desire to manipulate corporate media content. Television executives such as ABC's senior vice-president, Stephen Weiswasser, have made it clear that they do not want to 'turn passive consumers into active trollers on the Internet.'[10] Although Weiswasser made the comment in 1989, it more or less represents the attitude of the television industry today. A 2009 advertisement for NBC television proclaimed that the only place to see

NBC's programming online was at its official website. Not only was this statement incorrect, but most of my students and much of the Internet audience know it is untrue. There are many sites where one can find pirated television programming from all around the world. The television industry is in the same state of deep denial that the recording industry has been suffering from for over a decade. They have already lost the battle to contain the online audience and control their digital property.

In January 2008 YouTube started experimenting with a process that automatically muted the audio portion of any video that used a commercial soundtrack without permission. This has become a widespread practice that demonstrates the degree of control that YouTube has over amateur video production on its site. YouTube will develop many sophisticated mechanisms for automatically detecting copyright violations (and continue to steamroll over fair use in the process). What degree of threat this monitoring presents to amateur video production remains uncertain.

If YouTube engages in mass violations of fair use or too rigorously pursues copyright violations, it risks becoming the Napster of online video. Between 1999 and 2001 Napster was the single biggest site for downloading pirated music. After being shut down by court order, Napster was resurrected as a legitimate online music service. In this role, Napster was a minor player in the online music industry with modest revenues and was eventually sold to Best Buy for $121 million (U.S.) in 2008. Apple's online music service, iTunes, commands over 70 per cent of global online digital music sales, yet 95 per cent of downloaded music is pirated.[11] The moral of this story is that the online audience will migrate to services that provide the greatest amount of freedom. Contrary to the hopes of Rupert Murdoch and Time Warner, online content will continue to be free.

There is no guarantee that YouTube will not 'Napsterize' itself into insignificance. As a corporate entity, YouTube can severely limit the type of cultural production that may take place within its site. Yet YouTube and its corporate parent, Google, can do little to change the overall practices of amateur videographers across the Internet. YouTube's future state is unknown and its long-term financial success is a stock-market gamble. The world watched in awe throughout the latter half of 2008 as some of America's largest financial institutions degenerated into debt and dust. The same thing could happen in the computer industry. Microsoft, America Online, Google, YouTube – all are mortal.

YouTube is many things, but perhaps most of all it is a historical benchmark. It is a symbol of the Internet's transformation into a burgeoning medium for motion pictures – ours and theirs. It marks the globalization of amateur video practices. It is one more nail (an enormous spike) in the coffin of privacy. It has brought home movies out of the obscurity of our living rooms into the limelight of global media culture. It scares the hell out of Hollywood. It should scare the hell out of parents. It is the best thing to happen to advertisers in a long time. There is never enough room in one book to say all that could be said, so here I will conclude with an incomplete survey of the issues that YouTube faces and amateur online video presents as we move forward into the age of mass digital cultural production.

The Culture of Amateur Video

Can something so vast and varied as global digital amateur video practices (or even YouTube itself) be said to have characteristic cultural features? I think the answer is yes. Amateur videos are not simply representational practices. They are communicative, dialogic events that can provide the basis for community formation. As the foundation of this culture is the production of digital words and moving images, the nature of these digital bits will influence the nature of the culture. Unlike the culture of analogue television in the last century, a culture based on digital objects inherently is highly malleable. The culture of the Guttenberg press and commercial media worked by fixing meanings on static objects and privatizing shared stories via property laws. A culture based on digital storytelling is one wherein every story can be changed and altered. Also, digital stories that are valued by the community cannot be erased. The collective memory of digital culture is not easily subject to central powers and institutional controls.

YouTube videos can be seen in either low resolution or high resolution formats. While I have not discussed technical aspects of YouTube, one of the leading video theorists, Sean Cubitt, suggests that YouTube's choice of Adobe's .flv file format for videos 'mitigates against change … it limits us to the unchanging network of normal communication, and excludes us from what is beyond the frame and from change, it is art without hope.'[12] Video and film theory is filled with such instances of philosophical excess. Contrary to Cubitt, there is no reason to equate degrees of resolution or file formats with either banality or the sublime, either despair or hope. Such distinctions unnecessarily muddy the already less than pristine waters of new media theory. The culture of amateur online video is partially determined by the technological character of the medium, but that technology does not damn online video to a 'low culture' status of banal art without hope.

It is doubly odd that Cubitt should suggest that YouTube's videos mitigate against change, when we consider how amateur video practices differ from television. Within both YouTube and television one of the most prominent genres is reality-TV. Video diaries and many other genres of amateur videos are based on the urge to represent the reality of the situation. Yet when we look at commercial reality-TV, what we find is a product that is 'overwhelmingly conservative,' that closes 'down any collective impetus to effect change or to challenge the status quo.'[13] Largely because it is created by amateurs working in deinstitutionalized contexts, the same cannot be said about online amateur video. Television mitigates against change by providing a constant stream of propaganda on behalf of capitalism. Amateur online video has no such built-in conservative bias or structurally imposed ideology. Amateur video serves no master.

In 2006 media theorists looked into the depth of YouTube and cried foul – the 'Tube was guilty of lacking diversity.[14] Diversity is something that usually takes a while to develop in new Internet spaces and YouTube has been no exception. Now one must take a very narrow slice of YouTube to claim it lacks cultural diversity. When you wander through YouTube, you cannot help but realize that you are no longer in Kansas, Dorothy.

Authenticity is central to the culture of amateur online video and is a matter of great concern to YouTubers. Often videos acquire fame in an accidental manner. Many very widely viewed videos were spontaneous creations – a laughing baby, children being children, pets playing, a trip home from the dentist, a moment of musically inspired joy captured in front of a webcam. When individuals try to capitalize on their accidental fame and intentionally create a 'viral' video, the effort often falls flat, and the YouTube audience calls these individuals on their lack of authenticity.

Collective Cultural Production

Jenkins makes an intriguing suggestion that YouTube should have been called WeTube.[15] Amateur videos are often mashups that compile clips of other videos and build upon the creativity of others. Some of the most popular videos inspire an outburst of collective creativity. Parody, imitation, and commentary combine to turn a YouTube video into a pan-Internet media spectacle that frequently crosses over into the domain of television. Jean Burgess rightly suggests that popularity within YouTube is the result of a mass creative response to a video.[16] Whether we look at the *Numa Numa* lad, Charlie and Harry, BlackOleg's laughing baby, the song *Chocolate Rain*, or the performance *guitar*, what we find is that these amateur videos became famous because they inspired parody, imitation, and commentary. Fame within YouTube is a co-production involving many creators. YouTube works as a site because it supports the participatory culture of the larger Internet environment and its dominant mode of cultural collaboration. The fame it engenders cannot be attributed solely to the creative activity of a single individual.

Identity in Digital Culture

The effects of digital cultural production are contradictory and nowhere more so than in the realm of identity. Amateur video practices afford greater freedom for identity construction and also destabilize identity. Amateur video practices continually draw the past forward into the present. Charlie and Harry will always have the unaltered version of their home movies to look back upon. They will bring their past forward with them in a way that the masses have never quite been able to do. Yet they will also never be able to fully escape their past. They have lost control over some of their early childhood memories that have been altered and damaged by the YouTube community. Amateur video practices can lead to the destruction of self and identity.

Generation YouTube

In 2009 approximately 10 per cent of the Internet population created and uploaded amateur videos.[17] But Generation YouTube is being schooled in the art of blogging and video production. This is a generation that is growing up surrounded by digital cameras. This is a generation that is being born on YouTube itself. In all probability the rate of digital production among Generation YouTube will surge and far surpass today's volume of video uploads. The global audience for online video currently is estimated to total 250 million. By 2013 the number of Internet users who regularly access online video could quadruple and reach 1 billion.[18] Both the overall percentage of producers and consumers of online video are increasing dramatically. Generation YouTube one day may be as deeply engaged with online video as their grandparents were seduced by television.

Lifecasting

Lifecasting, also called webcasting, is a process in which individuals broadcast their life directly to the Internet, often wearing portable cameras. YouTube provides featured performers with live video streams that broadcast special events to the YouTube community. U2's Rose Bowl concert in October 2009 became YouTube's largest live streaming event when it generated 10 million streams across seven continents. The company is unlikely to extend webcasting to the entire community. Individuals have committed suicide while webcasting, and it is only a matter of time before someone broadcasts a real-time murder over the Internet.[19]

We will see more of lifecasting. I have tried a modest version with a home-made, portable 'headcam.' In the video *Dr. Strangelove Buys a Cup of Coffee* I go into a local coffee shop while wearing a digital camera duct-taped to the end of a stick that is also duct-taped to a bicycle helmet (there are many commercial varieties of helmet cameras). While YouTube may never allow lifecasting for the masses, many lifecasting events are recorded and uploaded to YouTube. Not quite the same thing, admittedly, but YouTube is a central repository for lifecasters.

A Search for Justice

As a public forum YouTube is used by those seeking redress for wrongs. Whether the misguided actions of a gold-digger or the plight of the Palestinians, we see individuals and groups on YouTube demanding justice and seeking to persuade the global community of the righteousness of their cause. In May 2008 Crystal Shinkle, a sixteen-year-old resident of Orlando, Florida, turned on her webcam in her bedroom and made a YouTube video wherein she complained that justice had been denied her.[20] Claiming that she had been raped and slighted by the justice system, Crystal took her case to the Internet. The Internet was not kind to her. Her background was further exposed, her privacy shredded, and her online identity destroyed by bloggers. Although her original video was quickly removed from her own YouTube channel, it is now found across the Internet and draws as many as 158,000 views on YouTube. The search for justice within the online community is a complicated affair.

The video camera may one day be a part of our clothing and capture most of our lives.

Over-Determination and the Defence of Piracy

Corporations want their media products to be widely viewed and discussed and need their celebrities to stay famous, yet they do not like to share. The current dominant strategy of commercial media is to disallow their videos to be embedded in unauthorized websites and the blogs of ordinary users. This is a mistake. They must learn to work with the most fundamental dynamics of the Internet and stop obsessing about control.

One of the more controversial claims in new media theory was written by Mark Poster: 'all citizens have an obligation to violate copyright law whenever they can.'[21] I'll not repeat Poster's rationale for this ethic but rather simply let it stand as an example of the emerging new media ethic of fair use. As Poster and many others have noted, media corporations engage in the most gross violation of fair use and use the law to overextend their rights, thus stifling creativity and undermining the collective good in the name of private profit.

In an article published in the *Wall Street Journal* titled 'In Defense of Piracy,' Lawrence Lessig comments on the case of Stephanie Lenz, who was threatened with absolute financial ruin by the brave and ethical lawyers of the Universal Music Group. The lawyers took exception to Lenz's outrageous act of videotaping her thirteen-month-old son, Holden, dancing to a song by Prince and posting it to YouTube (*'Let's Go Crazy' #1*). In the minds of the lawyers this post represented a dire loss of income to both Prince and Universal. Lessig describes the audio quality of the clip as 'terrible': 'No one would download Ms. Lenz's video to avoid paying Prince for his music. There was no plausible way in which Prince or Universal was being harmed by Holden Lenz.'[22] Nonetheless, Universal's lawyers pressed their demands and backed them up with a threat of a $150,000 (U.S.) fine. Whatever one might think about the ethics of digital piracy, it is quite obvious that corporations continue to act in a most unethical manner towards the active audience.

Lessig describes the current state of relations between the audience and corporations as hostile : 'We are in the middle of something of a war here.' The late and much reviled Jack Valenti referred to digital pirates as terrorists. As long as that attitude persists, corporations will merely fuel the flames of a battle that they cannot win. They can only criminalize mass creativity at the actual time when we need to think and act very differently towards the planet and each other. As Lessig suggests, we need to resist corporate claims – they threaten to corrupt the very rule of law – and we need to decriminalize Generation YouTube.

A Cultural Democracy?

Just as early proponents of the active audience wished to position television as a site of cultural (or semiotic) democracy, so too do some argue that video provides the grounds for a cultural democracy. In 1993 Cubitt cited video as a new dominant mode of visuality that had a 'uniquely democratic mission.'[23] Video practices promise to disrupt the mainstream representation of reality by injecting diversity into cyberspace and challenging commercial media to keep up or lose their audiences to the Internet. Cubitt puts his finger on the essence of amateur video's challenge to the reign of commercial media over the collective imagination when he notes that cultural activity is intensely policed.[24]

Yet we cannot celebrate YouTube and the Internet as pure, unfettered forms of cultural democracy. As Jenkins notes, YouTube 'lacks mechanisms which might encourage real diversity or the exchange of ideas. The Forums on YouTube are superficial at best and filled with hate speech at worst, meaning that anyone who tries to do work beyond the mainstream (however narrowly this is defined) is apt to face ridicule and harassment.'[25] Nonetheless, clearly we are looking at a mode of audience participation and cultural production that is far more participatory than television's very weak formation of cultural democracy. As José van Dijck notes, 'Cultural production can no longer be theorized exclusively in terms of industry or social stratification of consumers.'[26] The old label of 'consumer' oversimplifies the networked individual's position and role in society.

In 1996, three years after he speculated that video would provide the foundation for a new forum for cultural democracy, Cubitt reflected that whereas 'democracy and the internet seemed to be synonymous ... few people today would make that kind of statement.' Like many other media theorists, Cubitt saw the encroachment of commercial media into the Internet as the beginning of its normalization and transformation into yet another tool for commercial propaganda. Google's purchase of YouTube exemplifies the move 'towards the commercialization of cultural democracy ... There is no longer any reason to believe the internet is intrinsically democratic.'[27] The jury is still out on whether or not the Internet is intrinsically democratic. YouTube is many things, but it is not a democracy. It is a privately owned capitalist fiefdom. The minute we lose sight of that reality, we give ourselves over to YouTube's own propaganda.

The now famous cover of *Time* magazine proclaimed to the world, 'You control the Information Age. Welcome to your world,' and the cover story provided insight into capitalism's most central ideological notion.[28] Capitalism would have the mass audience believe that they are in control, that YouTube is their world, that the consumer is sovereign in the endless mall of America. As a spokesperson for corporate media, what *Time* did not say was most revealing – it did not tell readers that they are tenants, YouTube is the landlord and the village cop.

Beyond the confines of YouTube, amateur online video may be democratic, but perhaps the appropriate political metaphor for this collective media activity is anarchism. Democracy implies some form of collective to which the individual is held accountable. In its essence democracy is a form of collectively imposed accountability. On the Internet the amateur videographer is accountable to a nation or a collectivity only in the weakest sense. No one votes for what I point my camera at, and laws and market forces are very weak regulatory controls over my amateur media practices.

When we speak of the Internet as a democracy, we risk projecting our cherished political illusions onto this new realm. The simple fact that the dominant discourse about the Internet tends to use democracy as a preferred metaphor instead of some other form of political collectivity says much about the role of cultural assumptions in media theory. What people normally call democracy is not much more than a system that gives corporations vast control over ever increasing tracts of public resources and everyday life. YouTube may provide us with a means to bring the everyday further into the arena of political struggle. It may further subject the everyday to the logic of commodification. In all probability it will do both.

This Is Your Brand on YouTube

Every advance in the communicative possibilities that the Internet has enabled has increased the strength of the consumer's voice, and in this matter YouTube has been exemplary. In 2009 the Canadian musician Dave Carroll demonstrated the impact that YouTube can have on a company's reputation when United Airlines' baggage handlers broke his guitar. After the company refused to resolve the issue, Carroll told his tale in the YouTube music video *United Breaks Guitars*. After more than 6 million views and countless articles in the press, United finally compensated the Nova Scotian singer-songwriter and now uses the video to train personnel in customer relations. Carroll's video created enough bad public relations to cause United's stock price to dip by 10 per cent, costing shareholders approximately $180 million. Consumers have direct access to a potential international audience that measures in the millions. This means that companies can no longer count on problems being swept under the rug and disgruntled customers (or employees) giving up and going away. Your brand on YouTube may not look as good as your brand on television.

The Final Word?

What will happen to words in this age of the image? Mitchell Stephens suggests that 'literature in America and elsewhere continues to lose consequence.'[29] Much has been written about the supposed superior characteristics of the written word over video, and many fears have been expressed that we are coming to the end of the age of books. Without question, we are moving into a new age of mass participation in the creation and distribution of the image. Yet it is not quite clear that we are leaving words that far behind. A single video on YouTube can engender over 500,000 comments and more than 2.5 million words. YouTube may contain well over half a billion words, if not ten times that many. For every popular video there are thousands of blog entries discussing and analyzing the video. Alexandra Jusasz suggests that written thought on YouTube is subjected to a 'dumbing-down' because 'capacity to express ideas through words is almost entirely closed down on YouTube.'[30] This is an overstatement. In the video *The YouTube Community: A Commentary by Dr. Strangelove* I demonstrate that video can work as text. This video simply shows a text that scrolls as I read it. The entire text is also embedded 'beside' the video in the description area. Literature is not lost to video, it is simply re-presented.

Videos create words, engender dialogue, and foster interpersonal relations. Yet the outcome of this age of the image remains uncertain. All the words of the Enlightenment did not stop capitalism's ongoing destruction of the environment. All the images of the twentieth century have not brought an end to war, famine, grossly overpaid Wall Street executives, and other plagues upon humanity.

Finally, it seems appropriate to end a book about YouTube with a trip to YouTube. Grab your video camera; turn on your cellular phone, iPhone, or Blackberry; launch your webcam; make a video and put it on YouTube. We all like to watch.

Notes

Introduction

1 Gomes, 'Portals.'
2 Although not normally considered 'broadcast,' the structure and function of newspapers were and are comparable to television. Collectively, radio, television, and newspapers represent the holy trinity of corporate media that was responsible for many of the unique characteristics of industrial society in the twenty-first century.
3 Tartakoff, 'Analyst.'
4 Tanaka, 'Gabbing with Top Googler.' Seven months later the forecast for advertising and user-generated content remained largely unchanged. The media analysis company Deloitte likewise observed that 'advertisers are generally reluctant to place ads next to any [user-generated] content that could damage a client's brand.' Daswani, 'Deloitte Issues.'
5 Hardy and Hessel, 'GooTube.'
6 Tanaka, 'Gabbing with Top Googler.'
7 User-generated content certainly can be used to generate advertising revenue. The News Corporation's social networking site, MySpace, has 118 million users and generated almost $1 billion (U.S.) in revenue in 2007. According to journalist Brian Stelter, 'In the last few months, the bloom has come off social networking's rose. MySpace and its chief competitors, Facebook and Bebo, all have ambitious plans for making money but not enough proof that the plans are working.' 'MySpace Might Have Friends.' Also, by early 2009 Facebook still had failed to generate significant profits.
8 Tanaka, 'Gabbing with Top Googler.'
9 Learmonth, 'YouTube Moving the Needle.' Also see Brian Stelter, 'YouTube Videos.' Estimates for YouTube's annual advertising sales range from $120 million to $500 million (U.S.).

10 McIntyre, 'Introduction.'

11 Atton, *Alternative Internet*, x–xi.

12 Grossman, '*Time*'s Person of the Year.'

13 Also see Strangelove, 'Virtual Video Ethnography.' For an overview of the methodology and implications of digital ethnography, see Murthy, 'Digital Ethnography.'

14 Bailey, Guide to Qualitative Field Research, 2.

15 Burgess and Green, *YouTube*, 7.

16 By spring 2009 it was clear that a lack of standards in measurement was generating conflicting reports about online video viewership, but everyone agreed that video viewing was on the rise. Nonetheless, at this time it appeared that 'the video audience as a percentage of the overall Internet audience (at least in the U.S.) is pretty stable at around 70% to 80%.' Gannes, 'Online Video.' Also see Stelter, 'Hulu Questions Count.' Another study claims that self-reported data for television use is substantially under-reported, while online video and mobile video use is over-reported. Hess, 'Video Consumer Mapping Study.'

17 Junee, 'Zoinks!' In his blog, Junee notes: 'In mid-2007, six hours of video were uploaded to YouTube every minute. Then it grew to eight hours per minute, then 10, then 13. In January of this year, it became 15 hours of video uploaded every minute.'

18 Arrington, 'YouTube Video Streams.'

19 Hráček, 'Audiovizuální styl uživatelských YouTube videí.'

20 Worthen, 'Cisco Says.'

21 'Cisco Visual Networking Index.'

22 Sterling, 'YouTube Video.'

23 'Americans Viewed.'

24 'YouTube is …'

25 Gannes, 'Video Viewers Up 27%.'

26 Learmonth, 'Nielsen.'

27 Madden, 'Online Video,' iii.

28 Bourdieu, *Distinction*.

29 Madden, 'Online Video,' iii.

30 Poster, *Information Please*, 211. I am indebted to Poster for insight on the category of the everyday in the work of Henri Lefebvre.

31 Lefebvre, *Critique of Everyday Life*, 1:97.

32 Poster, *Information Please*, 213.

33 Lefebvre, *Critique of Everyday Life*, 92.

34 Poster, *Information Please*, 213.

35 Ibid., 230. Poster sees the everyday through the lens of an emerging 'post-human epoch.'

36 Wesch, *Anthropological introduction to YouTube*, video.
37 Lange, '(Mis)conceptions About YouTube,' 90.
38 'Survey Telecoms.'
39 Gantz, 'Diverse and Exploding Digital Universe,' 3.
40 Wesch, 'State of Research.'
41 Louderback, 'Where's the Outrage?'

1. Home Movies in a Global Village

 1 Pogue, 'Shazam!' In 1989 Czechoslovakian students recorded a revolution
    in progress and set up televisions on street corners to replay the VCR tapes.
    This suggests that a portable micro-projector may represent a substantial
    change not only in where videos are seen but in the political power and
    public impact of amateur videos. Here again we see the ever increasing loss
    of institutional control over the representation of events and moving im-
    ages.
 2 Zimmerman, 'Home Movie Movement,' 9.
 3 Ibid. Zimmerman offers a definition of amateur film from the perspective
    of Marxist theory: 'any work that operates outside of exchange values and is
    not produced to function as an exchange commodity.' Ibid.
 4 Trujillo, 'La Filmoteca de la Universidad Nacional Autónoma de México,'
    57.
 5 Ibid., 1.
 6 Keen, *Cult of the Amateur*, 5, 138.
 7 Kakutani, 'Cult of the Amateur.'
 8 Zimmerman, 'Home Movie Movement,' 5–6.
 9 Zimmerman, 'Geographies of Desire,' 85.
10 Zimmerman, *Reel Families*. See also Kattelle, *Home Movies*.
11 Chalfen, 'Home Movies as Cultural Documents,' 126.
12 Holliday, 'We've Been Framed.' Similarly, in 1988 Ben Singer bemoaned
    'the current lack of scholarly attention' that was being paid to the common
    phenomenon of home movie viewing. Fifteen years later Liz Czach would
    suggest, 'While significant contributions have been made into the study of
    home movies and amateur film, little has focused specifically on amateur
    video production.' Singer, 'Early Home Cinema,' 38. Czach, '*There's No Place
    Like Home Video*,' 114. For a critical response to Holliday see Sarah Pink,
    'More Visualising, More Methodologies.'
13 Cubitt, *Timeshift*, 16.
14 Zimmerman, 'Home Movie Movement,' xiii.
15 Homiak and Wintle, 'Human Studies Film Archives,' 42.
16 Cubitt, *Timeshift*, 2.

17 Geertz, *Local Knowledge*, 4, 16.
18 Although Robert J. Flaherty's *Nanook of the North* (1922) is often cited as the earliest full-length anthropological film, a recently discovered Finnish ethnographic film, *Arctic Pictures*, directed by Sakari Palsi was filmed in 1917. Cleverley, 'Ideology and Visual Anthropology,' 72.
19 Chalfen, 'Home Movies as Cultural Documents,' 128.
20 Worth, 'Margaret Mead,' 16.
21 Chalfen, 'Home Movies as Cultural Documents,' 129.
22 Ruoff, 'Home Movies of the Avant-Garde,' 297.
23 Jeremy Gibson quoted in Wayne, *Theorising Video Practice*, 65.
24 Zimmerman, 'Hollywood, Home Movies,' 23.
25 Wayne, *Theorising Video Practice*, 47.
26 YouTube vomit videos catch the attention of Christian Christensen in his article 'YouTube: The Evolution of Media?'
27 Kaplinsky-Dwarika, '"Most Annoying Sound."'
28 Chalfen, 'Home Movies as Cultural Documents,' 130, 131.
29 Ibid.
30 Scott, *American Politics in Hollywood Film*, 23.
31 Cubitt, *Timeshift*, 18.
32 Ibid., 19.
33 Chalfen, 'Home Movies as Cultural Documents,' 130.
34 Ibid., 132.
35 Poster, *Information Please*, 10, 22.
36 *Brother and Sister* has been deleted from YouTube and the channel no longer exists.
37 For example, see Pangea, *Emo Boy*. Here the videographer offers the YouTube audience the following explanation: 'This is my most controversial video and I don't want you to take it out of context even though there's no way you can't.'
38 Ruoff, 'Home Movies of the Avant-Garde,' 297.
39 Here Jeffrey Ruoff is paraphrasing Allan Sekula's argument on the contextual approach to photographic criticism found in *Photography Against the Grain*.
40 Ruoff, 'Home Movies of the Avant-Garde,' 297.
41 Chalfen, *Snapshot Versions of Life*.
42 Benjamin, 'Work of Art.'
43 Worth, 'Anthropological Politics of Symbolic Forms,' 346.
44 I am indebted to my brother, Steve Slade, for this phrase.
45 Cubitt, *Videography*, 3.
46 Ibid., 6, 7.

47 Both film studies and postmodern media theorists betray a curious fixation (dare I say fetish?) on Freudian and Lacainian psychoanalysis. On the role of psychoanalysis, see Flitterman-Lewis, 'Psychoanalysis, Film, and Television.'

48 Geertz, *Local Knowledge*, 163.

49 Trujillo, 'La Filmoteca de la Universidad Nacional Autónoma de México,' 57.

50 The videographer explained to me that this was part of a 'white trash' party he and his friends hold occasionally.

51 Taft, 'Men in Women's Clothes,' 131.

52 Bourdieu, *Distinction*, 57.

53 Geertz, *Interpretation of Cultures*, 14.

54 de Klerk, 'Home Away from Home,' 151.

55 I am indebted to Nico de Klerk for his insight on the thought of Eric de Kuyper, which de Klerk finds in 'Aux origines du cinéma,' 13.

56 Richard Chalfen suggests, quite rightly, that anthropologist David MacDougall overstated the case when he claimed that 'home movies tend to look similar in all societies.' As is often the case, early research into a cultural field often sees homogeneity where considerable heterogeneity exists. Claims of homogeneity are always problematic, but when one compares past domestic representation practices with those of the present, the differences are rather stark. MacDougall, 'Prospects of the Ethnographic Film,' 30.

57 Zimmerman, 'Hollywood, Home Movies,' 23.

58 Chalfen, 'Home Movies as Cultural Documents,' 128.

2. The Home and Family on YouTube

1 Grossman, 'Time's Person of the Year: You,' *Time*, 13 December 2006.

2 Moran, *No Place Like Home Video*, 28. Moran's fascinating study of home video, published four years before the creation of YouTube, does not directly offer an answer to the question 'What is video?' Neither will this study.

3 Burgess and Green, *YouTube*, 16.

4 Ibid.

5 Ibid., 28, 33.

6 Calavita, 'Futile Attempt at Medium Specificity.'

7 As an alternative to the techno-aesthetic paradigm that defines a medium by its technological properties, Moran suggests, 'at a particular time and in a local situation, a medium's specificity should be defined as those properties that lend themselves to the specific social functions and cultural agendas of vested individuals and groups ... Techno-aesthetic definitions lead inevita-

bly to determinist and essentialist prescriptions.' *No Place Like Home Video*, 31.

8 This analysis, conducted on 24 November 2008, excluded the videos *Sneezing Panda*, *Chocolate Rain*, and *My Webcam Tease* from the count, as their locations are indeterminate. An amateur animation, *Charlie the Unicorn*, was also excluded.

9 Moran, *No Place Like Home Video*, 48.

10 The key texts here are Chalfen, *Snapshot Versions of Life*, and Zimmerman, *Reel Families*.

11 Moran, *No Place Like Home Video*, 35. Herein I will not repeat the full scope of Moran's critique of older conceptions of home movies. My own analysis of YouTube and amateur online video is not meant to provide a thorough comparison of video and film or a complete critique of Zimmerman and Chalfen. Readers interested in a more in-depth analysis of these matters will find one in Moran, *No Place Like Home Video*.

12 Ibid. Moran also notes that both Zimmerman's and Chalfen's analyses of home movies 'broke significant ground' (ibid.).

13 Ibid.

14 RFC editor et al., 'RFC 2555.'

15 Darby, 'Technology.'

16 Rizzo, 'YouTube.'

17 Jenkins, 'From YouTube to WeTube.'

18 Brouwers, 'YouTube vs. O-Tube,' 117–18.

19 Moran, *No Place Like Home Video*, 105.

20 Home video has a complex relationship to television. Moran has argued that home-based practices of representation 'served as a primary influence on televisual codes of representation.' Ibid., 109.

21 Biressi and Nun, *Reality TV*.

22 Lange, 'Publicly Private and Privately Public.'

23 Again, I'll withhold the video title and the name of the videographer so as not to further his attempt to humiliate his sister. My thanks to Jamie Hammond for bringing this video to my attention.

24 Moran, *No Place Like Home Video*, 42, 43.

25 MacDonald, 'Teens: It's a Diary.'

26 Turnbull, 'Seven-Year-Old Bloggers.'

27 Solove, *Future of Reputation*, 17.

28 Gross and Acquisti, 'Information Revelation and Privacy.'

29 Solove, *Future of Reputation*, 17.

30 Damast, 'Admissions Office Finds Facebook.'

31 Comstock and Scharrer, *Media and the American Child*, 185–6.

32 Friedrich et al., 'Normative Sexual Behavior in Children.'

33 Martinson, *Sexual Life of Children*, 57. See also Hornor, 'Sexual Behavior in Children.'

34 See Juhasz, 'Learning From Fred'; Livingstone and Thumim, 'What is Fred Telling Us?'

35 Snelson, 'YouTube and Beyond.'

36 Harry, YouTube HDCYT channel description, 2009.

37 Ibid.

38 Wesch, *Anthropological introduction.*

39 Solove, *Future of Reputation*, 11.

40 Tice, 'Self-concept.'

41 Baumeister and Jones, 'When Self-Presentation Is Constrained.'

42 Tice, 'Self-concept,' 437, 449.

43 Cooley's theory of the looking glass self must be understood as a partial explanation. On this see Yeung and Martin, 'Looking Glass Self.'

44 There is a growing volume of evidence that Internet publishing activities such as blogging do affect social identity. I suspect that further research will demonstrate that amateur video will also be deeply implicated in identity processes. On the issue of blogging and identity see Pluempavarn and Panteli, 'Creation of Social Identity.'

45 Abrams, 'Social Identity, Social Cognition.'

46 Westin, *Privacy and Freedom*, 35.

47 Solove, *Future of Reputation*, 4.

48 Wallace, *Psychology of the Internet*, 34.

49 Wartella and Robb, 'Historical and Recurring Concerns.'

50 Nielson, 'Video Generation.'

51 Couldry, 'Liveness.' Couldry argues that our sense of liveness results not from a technology's essence or specificity but from materialized systems of classification.

52 Stone, 'Accuser.'

53 Livingstone, *Young People and the New Media*, 3.

54 Poster, *Information Please*, 171.

55 Meyrowitz, *No Sense of Place*, 238.

56 Poster, *Information Please*, 171.

57 Usha Goswami's review of child development literature leads to the conclusion that 'there is almost no empirical research relating to the question of the potentially harmful effects of the Internet and video games on child development.' 'Research Literature Review,' 51.

58 Wartella and Robb, 'Historical and Recurring Concerns,' 23.

59 Szabo, 'TV, Internet Harm Kids.'

60 Brown, Halpern, and L'Engle, 'Mass Media.'
61 Moran, *No Place Like Home Video*, 31.
62 Ibid., 41, 43.
63 Ibid., 41.

3. Video Diaries: The Real You in YouTube

1 Heffernan and Zeller, 'The Lonelygirl That Really Wasn't.'
2 Cubitt, *Timeshift*, 5.
3 Quoted in Young, 'An Anthropologist Explores,' A42.
4 Corner, *Art of Record*, 186.
5 Wharton, 'Click of the Mousse.'
6 Bakker, 'Tourism BC's Online Video Strategy.'
7 Jacobson, '"One-off Moments in Time."' Bambi Francisco Blog, 15 June 2006.
8 *Re: The Real Me.*
9 *The Real Me*, YouTube mangobanna channel.
10 *The Real Me*, YouTube lianeandthemusic channel.
11 *You Have No Idea Who I Am.*
12 Wright, 'Dare to Be Yourself.'
13 Evans, *Missing Persons*, 131.
14 Hobsbawm, *Age of Extremes*, 556.
15 Evans, *Missing Persons*, 131.
16 Stauffer, *Art of English Biography*, 55.
17 Landow, 'Autobiography, Autobiographicality.'
18 Nussbaum, 'Toward Conceptualizing Diary,' 136.
19 Ibid., 137.
20 Bolter, 'Digital Essentialism,' 196. Bolter also proposes that the search for the real in new media should be replaced with a quest for the transparent – an intriguing suggestion and one that is not without merit.
21 Ibid., 207.
22 Bolter and Grusin, *Remediation*.
23 The notion of an attention economy was first articulated by Herbert A. Simon in 'Designing Organizations.' Also see Goldhaber, 'Attention Economy and the Net.'
24 Quoted in Young, 'An Anthropologist Explores,' A42.
25 Liddell, *10 Confessions*. I could not hear all of Fiona's comments, so I contacted her via email for clarification.
26 Liddell, email message to author, 12 December 2008.

27 Renov, *Subject of Documentary*, 191, 206.

28 Foucault, *History of Sexuality*, 1:61–2.

29 Renov, *Subject of Documentary*, 93.

30 Foucault, *History of Sexuality*, 59.

31 White, *Tele-Advising*, 8.

32 Quoted in Levin, *Jean Rouch, Documentary Explorations*, 137. Also see Arthur, 'Moving Picture Cure.'

33 Renov, *Subject of Documentary*, 204.

34 There is an extensive body of literature on the presentation of the self in cyberspace, but scant attention has been paid to the specific issue of video diaries. See *Biography* 26, no. 1, for an early collection of articles examining Web-based (but not video) practices of autobiography. For an early study of the core/periphery structure of the video blogging community, see Warmbrodt, Sheng, and Hall, 'Social Network.' On the representation of the self in digital photography and online self-portraits see Walker, 'Mirrors and Shadows.'

35 Carter and Mankoff, 'Participants Do the Capturing.'

36 Rich et al., 'Video Intervention/prevention Assessment,' 156.

37 Gibson, 'Co-producing Video Diaries,' 5.

38 Bourdieu, 'Understanding,' 27.

39 Pini, 'Girls on Film.'

40 Ruby, 'Imagined Mirror,' 35. Ruby offers a rather rarified definition wherein 'reflexivity means that the producer deliberately and intentionally reveals to his audience the underlying epistemological assumptions that caused him to formulate a set of questions in a particular way' (35). In the end Ruby admits that such intentional reflexivity is actually rather hard to find in documentary forms, and he relies on a notion of accidental reflexivity (44). This difficulty may result from his use of an unnecessarily prescriptive notion of reflexivity.

41 On the reflexive character of both advertising and consumers, see Leiss et al., *Social Communication in Advertising*, 572.

42 Ruby, 'Imagined Mirror,' 44, 36.

43 Thompson, *Media and Modernity*.

44 Ruthlyn, *TheSargWP*.

45 Hausenblaus and Nack, 'Interactivity = Reflective Expressiveness.'

46 As of 30 January 2008 this video had garnered 8,472 views.

47 On the plural self see Rowan and Cooper, *Plural Self*. Given the multiplicity of theories on personality and selfhood, any one approach is likely to misstate the character of self-expression on the Internet. What is needed, and

what lies beyond the scope of my investigation of YouTube, is a framework of multiple theories that explores the multiplicity of selves that are revealed by online diarists.

48 David A. Huffaker and Sandra L. Calvert's analysis of online identity and language use among male and female teenager bloggers concludes that 'teenagers stay closer to reality in their online expressions of self than has previously been suggested.' This does not rule out the creative expression of the self within online environments. There is plenty of testimony by video bloggers about how they are both the same and not the same, which complicates this issue. More research is needed to see how the presentation of the self within video differs from written contexts such as Web logs. 'Gender, Identity, and Language Use.'

49 Egan, 'Encounters in Camera,' 596, 597.

50 Harley and Fitzpatrick, 'YouTube and Intergenerational Communication.'

51 Ibid.

52 Peter, *Re: Who are you.*

53 Peter, *corrupted*********by Zipster.*

54 Campbell and Harbord, 'Introduction,' 2.

55 Probyn, *Sexing the Self,* 91. Probyn provides an important correction to cultural theorists who would deny that no one can speak for another. She sees the 'eclipse of the category of experience' as arising from 'the enormous influence of structuralism and poststructuralism in cultural studies' (14).

56 Campbell and Harbord, *Temporalities, Autobiography,* 8.

57 Nick Couldry addresses the problems of certainty and scepticism in 'Speaking about Others and Speaking Personally.'

58 Evans sees autobiographers (who, I suggest, are more or less the same as YouTube's amateur video diarists) as struggling towards 'all-knowing information' (136). Yet this is an overstatement, which in turn generates unnecessary scepticism directed towards self-knowledge and self-representation. *Missing Persons,* 136, 143.

59 Ibid., 136, 142.

60 Renov, *Subject of Documentary,* 214.

61 Wiener and Rosenwald, 'A Moment's Monument,' 30.

62 Griffith, 'Looking for You,' 33.

63 Allport, *Use of Personal Documents.*

64 Wiener and Rosenwald, 'A Moment's Monument,' 56.

65 Ibid., 53.

66 Liddell, email message to the author, 14 November 2008.

67 Serfaty, 'Online Diaries,' 457.

68 *You Have No Idea Who I Am.*

69 Egan, 'Encounters in Camera,' 617.
70 Foucault, *History of Sexuality*, 61.
71 Dreyfus and Rabinow, *Michel Foucault*, 174–5.
72 Zalis, 'At Home in Cyberspace'; Zuern, 'Online Lives.'
73 Bruss, 'Eye for I,' 296. Bruss relies on a very minimalist and flawed definition of autobiography to arrive at her notion that the genre does not transfer to the filmic medium.
74 Renov, *Subject of Documentary*, 191–215.
75 Newman, 'Ze.'

4. Women of the 'Tube

1 Viewed 290 times (31 January 2009).
2 Rakow and Wackwitz, 'Voice in Feminist Communication Theory,' 102.
3 Nead, *Female Nude*, 68.
4 Jenna, *Benevolent Sexism and Compulsory Heterosexuality*. Viewed 4,507 times (31 January 2009). This video has since been deleted from YouTube.
5 Jenna, email to the author, 17 December 2008.
6 Valenti, *Friday Feminist Fuck You*. Viewed 53,280 times (31 January 2009).
7 Riotgirl93, reply to Jessica Valenti, *Friday Feminist Fuck You*.
8 Jenna, email to the author, 17 December 2008.
9 Molyneaux et al., 'Exploring the Gender Divide,' 6.
10 Ibid.
11 Ibid., 9
12 Smith and Watson, *Interfaces*, 3.
13 On the policing role of the male gaze, see ibid., 11n129.
14 While the male gaze is a theoretical concept steeped in psychoanalysis, here I am using it with its more everyday meaning of erotic objectification. For a more psychoanalytic use of the term, see Laura Mulvey's pioneering article, 'Visual Pleasure and Narrative Cinema.'
15 Murphy, 'Dialectical Gaze.'
16 Roach, *Stripping, Sex, and Popular Culture*.
17 MacKinnon, 'Sexuality, Pornography, and Method,' 140.
18 Williams, *Hard Core*, 4.
19 On the character of home-made pornography as art see Peraica, 'Chauvinist and Elitist Obstacles.'
20 Attwood, 'Sexed Up,' 77.
21 Juffer, *At Home with Pornography*, 227.
22 Lauren, *1/1: Meet Lauren*. Viewed 23,596 times (31 January 2009).
23 Kayley, *1/2: Meet Kayley*. Viewed 16,402 times (31 January 2009).

24 Hayley, *1/3: Meet Hayley*. Viewed 20,434 times (31 January 2009).

25 Kristina, *1/7: Meet Krisssstina!* Viewed 19,919 times (31 January 2009).

26 McKenna, Green, and Gleason, 'Relationship Formation on the Internet.'

27 Eaton, 'The YouTube Divorcée.'

28 Viewed 3,620,159 times (31 January 2009).

29 'Angry YouTube Divorcee Speaks Out.'

30 Steele, 'Web Watcher.'

31 Wright, *Becoming Black*, 230.

32 Michelle M. Wright argues in *Becoming Black* that the definition of Blackness suffers from 'insularity and oppressively homogeneous models' within literary discourses. No such homogeneity is found within the discourse on Blackness within YouTube. Ibid., 231.

33 For one of the more influential voices in the field of Black women and feminism, see hooks, *Ain't I a Woman*.

34 Rabbit, *I Got Weave In My Hair*. Viewed 6,400 times (31 January 2009).

35 Robin, *On Being Black* . Viewed 23,798 times (31 January 2009).

36 Kellner, *Media Culture*, 191.

37 Viewed 3,095 times (31 January 2009).

38 See, for example, Jacobs, Thomas, and Lang, *Two-Spirit People*; Garroutte, *Real Indians*.

39 Weissman, *Barbie*.

40 Karen, *Barbie Girl*. Viewed 8,136,612 times (31 January 2009).

41 Sapphi, *Happy! Crazy! Fun! FEMINISM*. Viewed 760 times (31 January 2009).

42 On *Sex and the City* and third-wave feminism, see Akass and McCabe, *Reading Sex and the City*.

43 *Ken & Barbie: Newlyweds*. Viewed 42, 236 times (31 January 2009).

44 Viewed 182 times (31 January 2009).

45 Bordo, *Unbearable Weight*, 24.

46 Cooper, 'What's Fat Activism?' 19.

47 Viewed 17,858 times (31 January 2009).

48 Mulveen, 'Interpretative Phenomenological Analysis.'

49 Heffernan, 'Narrow Minded.'

50 Diasi, 'Ana Sanctuary.'

51 Grosz, *Volatile Bodies*.

52 Ferreday, 'Unspeakable Bodies,' 277.

53 Brown and Tappan, 'Fighting Like a Girl.'

54 Quoted in Jones, 'Videotaped Florida Teen Beating.'

55 Quoted in Roberts, 'Police Stat Analysis.'

56 Spender, *Nattering on the Net*, 166.

57 Wylie, 'No Place for Women.'

58 For an overview of the complexity of women's participation online, see Scott, Semmens, and Willoughby, 'Women and the Internet.'

59 Fox, 'Older Americans and the Internet.'

60 'Gender Ratio of Netizens Becomes More Balanced,' WomenofChina.cn, 30 July 2008.

61 Among American Internet users who watched online videos in 2007, 31 per cent of young adult males used YouTube but only 22 per cent of females did so. I suspect that this gap will narrow over time. See Madden, 'Online Video.' The YouTube corporation claims that the overall gender ratio is 49 per cent female to 51 per cent male. 'YouTube is …'

62 Madden, 'Online Video'; Cheng, 'Nielsen.'

63 On the corporate feminization of the Internet versus the feminist uses of the Internet see Shade, 'Whose Global Knowledge?'

64 Alexa, reply to Duncan Riley, 'Women's Online Video Preferences Are Tamer Than Men's,' TechCrunch.com, 14 February 2008.

65 Quoted in Spicer and Taherreport, 'Girls and Young Women.'

66 Jenna, *Re: Haters, Vloggers*. Viewed 42 times (31 January 2009).

67 van Zoonen, 'Feminist Internet Studies,' 67.

68 Ibid.

69 Jenna, email to author, 17 December 2008.

70 Kennedy, 'Feminist Experiences in Cyberspace,' 707.

71 Travers, 'Parallel Subaltern Feminist Counterpublics.'

72 di Leonardo, *Gender at the Crossroads of Knowledge*, 190.

73 Ang, *Watching Dallas*, 119.

74 Allison, *Technology, Development, and Democracy*, 202, 207.

75 See, for example, Morahan, 'Women and Girls Last.' Martin Morahan claims, 'Gendered communication differences also affect Internet interactions and lead to male domination found in Internet discussion groups.' Yet this statement is contradicted by more recent research. Typical of this time and type of research, such claims rely on newsgroups and mailing list conversations from an earlier period of the Internet's history. Likewise, Carol R. Ronai, Barbara A. Zsembik, and Joe R. Feagin cite research from 1995 or earlier when they note in 1997, 'The Net is largely a masculine domain.' *Everyday Sexism in the Third Millennium*, 67.

76 Ang, *Watching Dallas*, 119. Also see John Fiske's analysis of soap operas as 'a feminine terrain where feminine meanings can be made and circulated.' *Television Culture*, 197.

77 Walkerdine, 'Femininity as Performance,' 57.

78 Carli and Bukatko, 'Gender, Communication, and Social Influence.'

79 Aries, *Men and Women in Interaction*, 6.

80 Ang, *Watching Dallas*, 14.
81 Viewed 15,508 times (31 January 2009).
82 Ang, *Watching Dallas*, 20.

5. The YouTube Community

1 See Watson, 'Why We Argue'; Driskell and Lyon, 'Are Virtual Communities True Communities?'
2 Bakardjieva, 'Virtual Togetherness,' 294.
3 Jones, 'Anthropology of YouTube.'
4 Forster, 'Psychological Sense of Community.' On why people join online communities, see Ridings and Gefen, 'Virtual Community Attraction.'
5 Silverman, 'Expanding Community,' 235.
6 Blanchard, 'Developing a Sense.'
7 Wilson and Peterson, 'Anthropology of Online Communities,' 450, 455.
8 Ibid., 455.
9 Ibid., 456.
10 Anderson, *Imagined Communities*, 6.
11 Wilson and Peterson, 'Anthropology of Online Communities,' 456–7.
12 Hampton and Wellman, 'Neighboring in Netville.'
13 Ortutay, 'Survey: Family Time Eroding.' Ortutay reports: 'Whether it's around the dinner table or just in front of the TV, U.S. families say they are spending less time together. The decline in family time coincides with a rise in Internet use and the popularity of social networks, though a new study stopped just short of assigning blame. The Annenberg Center for the Digital Future at the University of Southern California is reporting this week that 28 percent of Americans it interviewed last year said they have been spending less time with members of their households. That's nearly triple the 11 percent who said that in 2006. These people did not report spending less time with their friends, however.'
14 Quan-Haase and Wellman, 'Capitalizing on the Net,' 320. Also see Mohseni, Dowran, and Haghighat, 'Does the Internet Make People Socially Isolated?'
15 Jones, 'Internet and Its Social Landscape,' 17.
16 Fernback, 'Individual within the Collective,' 41.
17 Ibid., 42.
18 Douglas, *Risk and Blame*, 133.
19 Lucas Hilderbrand and Jean Burgess and Joshua Green have also noted this sense of entitlement among YouTubers. See Hilderbrand, 'Youtube'; Burgess and Green, 'Agency and Controversy.'
20 YouTube Team, 'YouTube for All of Us.'

21 *XXX PORN XXX*, YouTube UtubeAllFriends channel. At over 30 million views, this animation based on SimCity was once among YouTube's 100 most viewed videos.

22 Burgess and Green, 'Agency and Controversy,' 16.

23 YouTube Team, 'A YouTube for All of Us.'

24 The tags on this video are 'PORN, anal, sex, oral, ass, tit, fuck, get fucked, cunt, dick, pussy, porno, fucking, naked, lingerie, horny, milf, hot, mom, babe, fetish.'

25 Quoted in San Miguel, 'YouTube Shines Spotlight.'

26 Nufff, *Algorithmically Demote This!*

27 Burgess and Green, 'Agency and Controversy,' 5.

28 Ibid., 2.

29 The largest unresolved lawsuit faced by YouTube is a claim for $1 billion in damages by the entertainment corporation Viacom. For an overview and analysis of the *Viacom v. YouTube* litigation, see O'Brien, 'Copyright Challenges.' O'Brien suggests that 'there are certainly strong arguments in favour of Viacom' (233).

30 Holson, 'Hollywood Asks YouTube.'

31 For an overview of the legal issues surrounding the DMCA, see George and Scerri, 'Web 2.0 and User-Generated Content'; Sithigh, 'Mass Age of Internet Law'; Lastowka and Dougherty, 'Copyright.'

32 On the current state of the information wars between citizens and governments, see Faris, Wang, and Palfrey, 'Censorship 2.0.'

33 On the range of copyright knowledge among YouTubers, see Aufderheide, Jaszi, and Brown, 'The Good, the Bad, and the Confusing.'

34 Ibid., 16.

35 Lunceford and Lunceford, 'Meh. The Irrelevance of Copyright,' 33.

36 Dames, 'Copyright and the Speed Limit,' 18.

37 Brown, 'Copyright Infringement,' 16.

38 Gillespie, *Wired Shut*, 10.

39 Hunt, 'Copyright and YouTube,' 203. Likewise, Patricia Aufderheide and Peter Jaszi argue that a substantial amount of copyrighted material included in amateur video should be permissible under the fair use provisions of copyright law. 'Recut, Reframe, Recycle.' Offering a contrary opinion, Parul Kumar argues that YouTube cannot shield itself by using the fair use defence. 'Locating the Boundary.'

40 Brouwers, 'YouTube vs O-Tube,' 107.

41 This trend of negative comments tends to reverse among more recent written comments. Given that Oprah's first video addressing the YouTube community was made in November 2007, it may be the case that she initially

attracted a wide audience and considerable criticism for her video as one of the first major celebrities to use YouTube. The more recent viewings of her first video may reflect a narrower Oprah fan audience and thus less hostile commentary.

42 Chin and Hills, 'Restricted Confessions?' 253.
43 Brouwers, 'YouTube vs O-Tube,' 113.
44 Ibid., 117.
45 Burgess and Green, 'Agency and Controversy,' 9.
46 Itzkoff, 'One-Way Ticket to Disaster.'
47 See written comments on Crocker, *Chris Crocker – LEAVE BRITNEY ALONE!*
48 On the role of parody and inside jokes within YouTube fan videos, see Lamerichs, 'It's a Small World.'
49 Ibid., 64.
50 Jenkins, *Convergence Culture*, 244.
51 Karpovich, 'Reframing Fan Videos,' 17.
52 Baym, 'New Shape of Online Community.' *First Monday* 12 (August 2007).
53 Moon, 'Dance Dance Evolution,' 1–3.
54 Tannen, *Argument Culture*, 64.
55 Wood, *Bee in the Mouth*, 26–7.
56 Quoted in Rivlin, 'Hate Messages on Google.'
57 Comment seen attached to *Top 10 Racist Jewish outbursts*, YouTube shamraiz channel, 11 November 2006.
58 Julian, *I Hate Black People Because …*
59 For one of the earliest analyses of violence on YouTube, see Kambouri and Hatzopoulos, 'Making Violent Practices Public.'
60 Colombani, 'We Are All Americans.'
61 Lange, 'Commenting on Comments,' 1.
62 Kampman, 'Flagging or Fagging,' 160.
63 Benevenuto et al., 'Identifying Video Spammers.'
64 Burgess and Green, 'Agency and Controversy,' 6.
65 Ibid., 13, 49.
66 Ibid., 14.
67 Benevenuto et al., 'Characterizing Video Responses,' 1.
68 Paolillo, 'Structure and Network,' 1.
69 Lange, 'Publicly Private and Privately Public,' 3, 8.
70 Burgess and Green, *YouTube*, 103.
71 Hill, *Fan Cultures*, 65–89.
72 See Clifford and Marcus, *Writing Culture*.
73 Rabinow, 'Representations Are Social Facts,' 250.

74 Paolillo, 'Structure and Network,' 8.

75 Ibid.

76 Cheng, Dale, and Liu, 'Internet Short Video Sharing,' 3.

77 The following study shows that '80% of videos requested on a given day are older than 1 month.' Cha et al., 'I Tube, You Tube,' 7.

78 Landry and Guzdial, 'Art or Circus?'

79 Kruitbosch and Nack, 'Broadcast Yourself on YouTube?'

80 Cha et al., 'I Tube, You Tube,' 8–12.

81 Cheng, 'Internet Short Video Sharing,' 1.

82 For example, see Lee, 'Locality Aware Peer Assisted Delivery'; Leighton, 'Improving Performance on the Internet.'

83 Horrigan, 'Home Broadband Adoption 2008.'

84 Gunning, 'Cinema of Attractions,' 58.

85 Rizzo, 'YouTube.'

86 Gunning, 'Cinema of Attractions,' 58.

87 Rizzo, 'YouTube.'

88 Galbi, 'Stories Largely Missing.'

89 De Vos, *Storytelling for Young Adults*, 16; Block, *Visual Story*, 173.

90 On the definition of story see Rayfield, 'What Is a Story?'

91 'Star Wars Kid.'

92 The most viewed viral video is footage of Ghyslain Raza, of Quebec, Canada, who was the subject of the famous *Star Wars Kid* amateur video, which has been viewed over 900 million times. As this video predates YouTube by three years and has been acknowledged within the academic record, it is not discussed herein. The original *Star Wars Kid* video is on a YouTube channel attributed (but unconfirmed) to Ghyslain Raza. Many of the 55,000 comments on the video are hostile and demeaning.

93 Feuer and George, 'Internet Fame Is Cruel Mistress.'

94 Wortham, 'Cheating Fans.'

95 Rainie and Madden, 'Web Survey of Musicians and Songwriters,' 3.

96 Quoted in Battelle, 'Brief Interview with Michael Wesch.'

97 Ibid.

98 Innis, *Bias of Communication.*

99 Carpenter, 'New Languages,' 173.

6. The YouTube Wars: Politics, Religion, and Armed Conflict

1 Whittaker, Issacs, and O'Day, 'Widening the Net.'

2 Carlson and Strandberg, 'Riding the Web 2.0 Wave,' 2.

3 Smith and Rainie, 'Internet and the 2008 Election.'
4 'I remain somewhat uncertain regarding the role of voter-generated content in shaping political discourse.' Tryon, '"Why 2008 Won't Be Like 1984."'
5 Rainie, 'E-Citizen Planet,' 2.
6 Gueorguieva, 'Voters, MySpace, and YouTube.' Also see Walters, 'Online Stump,' 60.
7 'First YouTube Election.'
8 Schejter and Yemini, '"Justice, and Only Justice,"' 173.
9 Ruffini, 'The Internet is TV.'
10 Hart, 'On YouTube.'
11 Martin, 'Don Martin's Stories.'
12 Kiley, 'YouTube Election Ratifies.'
13 Turkheimer, 'YouTube Moment in Politics,' 98.
14 Dickinson, 'First YouTube Election.'
15 Castells, 'Communication, Power and Counter-Power,' 255.
16 Quoted in Lizza, 'YouTube Election.'
17 Fairbanks, 'YouTube Election.'
18 Gumbel, '"YouTube Elections."'
19 McArthur, 'YouTube No Place.'
20 Carlson and Strandberg, 'Riding the Web 2.0 Wave,' 15, 23.
21 Macnamara, 'E-Electioneering,' 6.
22 Milliken et al., 'User-Generated Online Video.'
23 This video appears to be inspired by an earlier YouTube video, *Adolph Harper loses it.*
24 Wallsten, '"Yes We Can."'
25 Moscovitch, 'Hitler's Downfall, Parodied.'
26 This video was made by a YouTube member who purports to be an 'award-winning Canadian documentary filmmaker and screenwriter who prefers to remain anonymous.' This raises the question, 'can a media professional make an amateur video?'
27 Richler, 'Canadian Year in Video.'
28 Van Dijk, *Discourse and Structures of Power.*
29 Hartley, 'YouTube, Digital ,' 11, 15.
30 Castells, 'Communication, Power and Counter-Power,' 258–9.
31 Slabbert, 'Orwell's Ghost,' 359.
32 Ginsberg, *Captive Public.*
33 Hauser, *Vernacular Voices*, 109, 89.
34 Shah, 'YouTube Crawling.'
35 Hunt, 'Obscenity and the Origins of Modernity.'
36 Lin and Hauptmann, 'Identifying Ideological Perspectives.'

37 Gare, 'Politics of Recognition.'
38 Berlet, 'When Hate Went Online.'
39 'German Jewish Group.'
40 Strangelove, 'Redefining the Limits to Thought.'
41 Seltzer, 'DMCA "Repeat Infringers."'
42 Ortega, 'What to Get L. Ron Hubbard.'
43 Metz, 'YouTube Rolls Out.'
44 Aleo-Carreira, 'Bogus Anti-Scientology DMCA Notices.'
45 Oksanen and Välimäki, 'Theory of Deterrence.'
46 Dawson and Hennebry, 'New Religions and the Internet,' 283.
47 Beschizza, 'Creationist vs. Atheist.'
48 Heffernan, 'Many Tribes of YouTube.'
49 Berman, Nelson, and Yiu, 'Blasphemy Challenge.'
50 Quoted in Le Beau, *The Atheist*, 290.
51 Ortega, 'What to Get L. Ron Hubbard.'
52 Cohen, *There Is a War.*
53 *Evolution of Jihad Video*, 6–8.
54 'Dark Web Terrorism Research.'
55 Associated Press, 'Egyptian Student Gets 15 Years.'
56 Chen, Thoms, and Fu, 'Cyber Extremism in Web 2.0,' 5. Also see Salem,
     Reid, and Chen, 'Jihadi Extremist Groups' Videos.'
57 Quoted in 'YouTube Contributes,' 5.
58 Shaidle, 'Facebook Jihad.'
59 Rid, 'Bundeswehr's New Media Challenge,' 105.
60 'Israel Clashes with YouTube.'
61 Butters, 'Fighting the Media War.' This sentiment was echoed by Chottiner,
     'YouTube War.' Also see Baker, 'Israel Gains in Media Blitz.'
62 Minty, 'Waging the Web Wars.'
63 Gissin, 'No Israeli Apology.'
64 Butters, 'Fighting the Media War.'
65 Cox, 'YouTube War.'
66 Christensen, 'Uploading Dissonance,' 155.
67 Losh, 'Government YouTube.'
68 Carruthers, 'No One's Looking,' 70.
69 Ibid., 73.
70 Blanton, 'Poll.'
71 Carruthers, 'No One's Looking,' 75.
72 Naím, 'YouTube Effect.'
73 A further indication of the intensity of political debate that occurs within
     YouTube is seen in the amateur video *Macedonia is Greece.* Viewed 3 million

times, this video has garnered over 700,000 written replies and is the most commented upon YouTube video.

74 Beschizza, 'Creationist vs. Atheist.'

75 Seitz, 'Copy Rites.'

76 When Ottawa's city buses went on strike during Christmas 2008 and on into the new year, local citizens used YouTube to express their opinions on the matter. The *Downfall* clip was turned into a parody wherein the strike ruins Hitler's plans for a weekend trip to Ottawa (*OC Transpo Strike Ruins Hitler's Plans*). Amateur videographers captured action at the picket lines and union members made videos that argued strike positions.

7. The Post-Television Audience

1 Williams, *Television, Technology and Cultural Form*, 30.

2 Curran, *Media and Power*, 116.

3 Curran, Morley, and Walkerdine, *Cultural Studies and Communication*, 109. Along with Walkerdine, Curran also sees a relationship between Fiske's celebration of the audience's interpretative power and the general context of American individualism. Thus, Fiske's notion of an autonomous cultural economy is 'not very different from the American liberal tradition in which the media are analysed in isolation from power relationships or are situated within a model of society in which, it is assumed, power is widely diffused' (259–60).

4 Fiske, *Television Culture*, 316.

5 Fiske, *Understanding Popular Culture*, 20–1.

6 Wolf, *Pathways of Power*, 379–82.

7 Curran, Morley, and Walkerdine, *Cultural Studies and Communication*, 268.

8 Strangelove, *Empire of Mind*.

9 Ang, *Living Room Wars*, 8.

10 Heffernan, 'Hitler Meme.'

11 Viewed 17,554 times (5 February 2009).

12 Heffernan, 'Hitler Meme.'

13 Fiske, *Power Plays, Power Works*.

14 Ang, *Living Room Wars*, 9.

15 Curran, *Media and Power*, 165.

16 Ang, *Living Room Wars*, 12, 13, 15.

17 Barlow, 'Declaration of Independence.'

18 Ashcroft and Griffins, *Post-Colonial Studies Reader*, 2.

19 Ibid.

20 Lovink, 'Radical Media Pragmatism Strategies,' 33.

21 Gilder, *Life After Television*, 206–7.

22 Agre, 'Cyberspace as American Culture.'

23 Gilder, *Life After Television*, 49.

24 Lenert, 'Communication Theory Perspective.'

25 Cubitt, *Videography*, 201.

26 Jancovich and Faire, *Place of the Audience*, 226.

27 Napoli, 'Audience Measurement and Media Policy,' 27.

28 Elliot, 'Media Business.'

29 Saucier, 'New CBC.'

30 Zoglin and Harbison, 'Goodbye to the Mass Audience.'

31 Lebo, 'First Release of Findings.'

32 Bond, 'Online Video Chips Away.'

33 Nielsen, 'TV Viewing Among Kids' and 'Americans Watching More TV.'

34 Quoted in Flint and Kelly, 'New Signs of Strain.'

35 Peers, 'Internet Killed the Video Star.'

36 Stelter and Stone, 'Digital Pirates Winning Battle.'

37 Gonsalves, 'Young TV Viewers.'

38 Quoted in Waxman, 'Hollywood 2.0 with Barry Diller.'

39 Stone and Vance, '$200 Laptops.'

40 Keating, 'U.S. Woes.'

41 Pegoraro, 'TV Over the Web.'

42 Shirky, *Here Comes Everybody.*

43 Quoted in Albanesius, 'Online Video No Threat.'

44 Van Dijck, 'Television 2.0.'

45 Wesch, *Anthropological introduction.*

46 Moran, *No Place Like Home Video*, 97. Cubitt likewise describes television not as a separate practice but as 'the pinnacle of an institutional and discursive hierarchy of video practices.' *Videography*, 134.

47 Cubitt, *Videography*, 32.

48 Moran, *No Place Like Home Video*, 100, 101.

49 Moran cites Douglas Davis as an example of 'an unwarranted ideology of presence' that attempts to define video as more live, transparent, and reality based than television (104). See Davis, *Five Myths of Television.*

50 Moran, *No Place Like Home Video*, 104.

51 Adorno, 'Culture Industry Reconsidered.'

52 Adorno and Horkheimer, *Dialectic of Enlightenment*, 120.

53 Gunster, *Capitalizing on Culture*, 277.

54 Atton, *Alternative Internet*, vi.

55 Klinger, *Beyond the Multiplex.*

56 Atton, *Alternative Internet*, 26, 46.

57 Oulliette, 'Camcorder Dos and Don'ts.'
58 Atton, *Alternative Internet*, 159.
59 On the accusation that corporate media is a 'significantly anti-democratic force' see McChesney, *Problem of the Media*, 18.
60 Atton, *Alternative Internet*, 158.
61 Rodriguez, *Fissures in the Mediascape.*
62 Atton, *Alternative Internet*, 258.
63 Like so many other industries caught up in the unprecedented financial turmoil of these current times, television may succumb to the mightiest force in capitalism – the consumer's wallet. In 2008 the results of a Jupiter-Research study suggested that in a troubled economy consumers would stop paying for movie tickets and premium cable channels long before they disconnected their broadband Internet accounts. See Barnes, 'For a Thrifty Audience.'
64 Atton, *Alternative Internet*, 105.

Conclusion

1 This quote has been variously attributed to Aristotle, the Hopi, the Navajo, and Plato. In the *Republic*, Book II, Plato says: 'Then the first thing will be to establish a censorship of the writers of fiction, and let the censors receive any tale of fiction which is good, and reject the bad; and we will desire mothers and nurses to tell their children the authorised ones only. Let them fashion the mind with such tales, even more fondly than they mould the body with their hands; but most of those which are now in use must be discarded' (49). Special thanks to my father-in-law, Raymond St. Jacques, for his assistance here.
2 Smith, 'Analysis.'
3 Manovich, 'Practice of Everyday (Media) Life,' 36.
4 Terranova, 'Free Labor.'
5 YouTube, 'Terms of Use.'
6 Jenkins, 'Nine Propositions.'
7 Schäfer, 'Bastard Culture,' 291.
8 Noam, 'Economics of User Generated Content.'
9 Lessig, *Remix.*
10 Quoted in Kelly, 'We Are the Web.'
11 Rogers, 'Piracy Still Prevailing.'
12 Cubitt, 'Codecs and Capability,' 49.
13 Biressi and Nunn, *Reality TV*, 155.
14 McMurria, 'YouTube Community.'

15 Jenkins, 'From YouTube to WeTube ...'

16 Burgess, '"Is Your Chocolate Rain."'

17 This is at best a rough estimate, largely based on Madden's 2007 analysis of American online video users. No concrete global figures are available. Madden established that 57 per cent of Americans viewed online video and 13 per cent of this group uploaded videos. That would put the number of American Internet users who upload videos at approximately 10 per cent of the entire American Internet community. America lags behind many other nations in terms of high bandwidth Internet users. High-bandwidth Internet users more frequently upload multimedia content. Thus, a reasonable estimate for global participation in online video uploading is around 10 per cent. No estimates for the total number of videos uploaded daily across the globe are available. Given that the YouTube community uploads over 300,000 videos daily, the global total across all Internet sites could measure anywhere from 500,000 to 2 million videos per day. Madden, 'Online Video.' In 2007 Susan Faulkner and Jay Melican also suggested that 10 per cent of Internet users are uploading video to the Web. 'Getting Noticed, Showing-Off,' 46.

18 'More Than One Billion Users.'

19 Stelter, 'Web Suicide Viewed Live.'

20 Frantz, 'Teen Alleging Rape.'

21 Poster, *Information Please*, 198.

22 Lessig, 'In Defense of Piracy.'

23 Cubitt, *Videography*, 202.

24 Cubitt, *Timeshift*, 6.

25 Jenkins, 'From YouTube to WeTube ...'

26 Van Dijck, 'Users Like You?' 53.

27 Mills, 'Sean Cubitt.'

28 *Time*, 13 December 2006.

29 Stephens, *Rise of the Image*, 207.

30 Jusasz, 'Why Not (to) Teach,' 136.

YouTube Videos

Adolph Harper loses it. YouTube directorpictures channel, 20 August 2008.

Bahner, Adam Nyerere. *'Chocolate Rain' Original Song by Tay Zonday.* YouTube TayZonday channel, 22 April 2007.

Barack Obama masturbating. YouTube wavinginavacuum channel, 3 January 2009.

Barbie Girl dance. YouTube tapjazz95 channel. 29 April 2008. Removed from YouTube.

Bree. *First Blog / Dorkiness Prevails,* YouTube Lonelygirl15 channel, 16 June 2006.

Brolsma, Gary. *New Numa – The Return of Gary Brolsma!* YouTube NewNuma channel, 8 September 2008.

Brother and Sister. YouTube volleyballhottie09 channel, 9 September 2007.

Buckley, Michael. *The Real Me.* YouTube whatthebuckshow channel, 8 May 2008.

Calin. *important message for everyone.* YouTube gruia12345 channel, 28 October 2008.

Carroll, Dave. *United Breaks Guitars.* YouTube sonsofmaxwell channel, 6 July 2009.

Caryn. *andsanp is gone?* YouTube carynis1hotassgirl channel, 13 March 2008.

Claire. *Unassisted Birth.* YouTube sociallyskilled channel, 22 September 2007.

Compagucci, Melissa Jenna. *Re: Being a Chick on YouTube,* YouTube melissajenna channel, 8 August 2007.

Crissy. *RE: Why I youtube.* YouTube PhunkyPrincess channel, 1 December 2007.

Crocker, Chris. *Chris Crocker – Individuality/Gender.* YouTube itschriscrocker channel, 12 August 2007.

– *Chris Crocker – LEAVE BRITNEY ALONE!* YouTube itschriscrocker channel, 10 September 2007.

Current, Edward. *The Atheist Delusion.* YouTube EdwardCurrent channel, 2 June 2007.

Delilah. *Black Women Dropped the Ball (Attitude, Weave and Disrespect)*. YouTube melosidad1 channel, 17 March 2008.

Disneyland Home Movies. YouTube WEBmikey channel, 7 January 2007.

Dyer, Miles. *How to make your 1st Vlog! START TODAY!* YouTube blade376 channel, 24 March 2007.

Emily. *barbie and the city (mature audiences only)*. YouTube ohemilygee channel, 2 November 2006.

Feminist. *This is What a Feminist Looks Like*. YouTube feministmajority channel, 22 March 2008.

Giles, Sharron Rosa. *Meet My Parents …*' YouTube charityrose2 channel, 30 October 2007.

Hahaha. YouTube BlackOleg channel, 1 November 2006.

Hail Hitler, Hail Obama. YouTube ATLAHWorldWide channel, 9 June 2008.

Happy New Year 2008. YouTube FlimFlamFilms channel, 31 December 2008.

Harry. *Charlie bit my finger – again!* YouTube HDCYT channel, 22 May 2007.

Hayley. *1/3: Meet Hayley*. YouTube fiveawesomegirls channel, 3 January 2008.

Huge Black Girl Fight. YouTube botepipapopote channel, 5 March 2008.

Jenna. *Benevolent Sexism and Compulsory Heterosexuality*. YouTube jennaow channel, 11 January 2008.

Jenna. *Re: Haters, Vloggers, Respect, and YouTube*. YouTube gothicgirl0666 channel, 2 May 2008.

Jimmy. *The Real Me*. YouTube ledfan38 channel, 14 May 2008.

Julian. *I Hate Black People Because …* YouTube davispomo channel, 8 November 2006.

Karen. *Barbie Girl: 5th grade talent show*. YouTube ckfools channel, 7 June 2006.

Kayley. *1/2: Meet Kayley*. YouTube fiveawesomegirls channel, 2 January 2008.

Ken & Barbie: Newlyweds. YouTube Suedehead237 channel, 22 August 2007.

Kristina. *1/7: Meet Krisssstina!* YouTube fiveawesomegirls channel, 7 January 2008.

Lauren. *1/1: Meet Lauren*. YouTube fiveawesomegirls channel, 1 January 2008.

Lauren. *3/25: The Real Lauren*. YouTube fiveawesomegirls channel, 25 March 2005.

Liddell, Fiona. *10 Confessions*. YouTube FIONAshizz channel, 4 June 2008.

Macedonia is Greece, YouTube makedoniatruth channel, 13 January 2007.

Misha. *Native American Women*. YouTube dadasopher channel, 5 July 2008.

Morgan. *Is YouTube a Feminist Space?* YouTube msross06 channel, 15 October 2008.

Nash, Joy. *A Fat Rant*. YouTube joynash1 channel, 17 March 2007.

Nufff. *Algorithmically Demote This!* YouTube NufffRespect channel, 18 December 2008.

OMG I Am Fat (Like I didn't Already Know). YouTube DeJadela channel, 28 November 2006.

Oral. YouTube kicesie channel, 2 April 2007.

Pangea, Matthew David. *Emo Boy.* YouTube pangeaforever channel, 27 December 2007.

Paulina. *Not Barbie Dolls.* YouTube Bubelcoyot channel, 5 April 2008.

Peace for Iraq – Stop the war. YouTube davybrasco channel, 15 February 2007.

Peter. *corrupted********by Zipster.* YouTube geriatric1927 channel, 7 December 2008.

– *MERRY XMAS.* YouTube geriatric1927 channel, 11 December 2008.

– *Re: Who are you ... Who, Who ... Who, Who.* YouTube geriatric1927 channel, 8 August 2006.

QueenJean. *He's gonna kick MY ass?* YouTube thesugarmama channel, 27 March 2007.

Quigley, Adam. *How does the world feel to you through YouTube?* YouTube quigleyadam channel, 10 November 2008.

Rabbit. *I Got Weave In My Hair So Must Wanna B White.* YouTube chynadoll21388 channel, 9 June 2008.

Re: Haircut Experience Update!!! SHAVE MY HEAD, YouTube NorCaLg33k channel, 17 April 2007.

Response to a response to a video :D. YouTube MsDiscord channel, 11 June 2008.

Re: The Real Me. YouTube InfoDissemination channel, 13 May 2008.

Robin. *On Being Black and a Woman (Part 1 of 2).* YouTube spokenlife channel, 23 April 2007.

Ruthlyn. *TheSargWP – You Used Confliction Out Of Context, Sweetie!* YouTube Ruthlyn channel, 24 November 2008.

Sapphi. *Happy! Crazy! Fun! FEMINISM.* YouTube loveSAPPHI channel, 20 October 2007.

Sarah. *Pro Ana.* YouTube pleasetakemeana44 channel, 25 March 2007.

Sasha CHEER for Team U.S.A!!! YouTube bankassociatecheer, 12 June 2008.

Shel. *the real me.* YouTube NannyCam channel, 13 May 2008.

Smith, Gareth. *Re: The Real Me.* YouTube Redzool channel, 13 May 2008.

Smooth, Jay. *Vlogging is Stupid.* YouTube illdoc1 channel, 1 September 2008.

Stephane Dion: White and Nerdy. YouTube doctorwho4ever channel, 2 January 2007.

Strangelove, Michael. *Dr. Strangelove Unboxing an Easel.* YouTube empireofmind channel, 18 May 2008.

– *Dr. Strangelove Smashes a Cell Phone.* YouTube empireofmind channel, 20 October 2008.

- *The Real in the HyperReal (A Reply to: Web 2.0 … The Machine is Us/ing Us)*. YouTube empireofmind channel, 13 February 2007.
- *The YouTube Community: A Commentary by Dr. Strangelove*. YouTube empireofmind channel, 5 February 2009.

That Damn Downfall Clip. YouTube TheDarwinian channel, 29 January 2007.

The Blasphemy Challenge. YouTube BlasphemyChallenge channel, 7 December 2006.

The Harper Dictatorship. YouTube harperdictatorship channel, 4 December 2008.

The Real Me. YouTube lianeandthemusic channel, 26 March 2008.

The Real Me. YouTube mangobanna channel, 29 March 2008.

The Real Me. YouTube THEFRANKSHOW channel, 13 May 2008.

Thinspiration! YouTube Miyavismxdesux channel, 26 May 2007.

Thunderstruck – US Army. YouTube thatdood001 channel, 22 August 2006.

Tom Cruise Scientology Video – (Original Uncut). YouTube Aleteuk channel, 17 January 2008.

Valenti, Jessica. *Friday Feminist Fuck You: Online Misogyny (YouTube edition)*. YouTube Feministing channel, 30 May 2008.

Walsh-Smith, Tricia. *Tricia Walsh Smith – The video that started it all!* YouTube walshsmith1 channel, 10 April 2008.

Wesch, Michael. *An anthropological introduction to YouTube*. YouTube mwesch channel, 26 July 2008.

- *Web 2.0 … The Machine is Us/ing Us*, YouTube mwesch channel, 31 January 2007.

White Trash Weddin – what the Hell's a YouTube? YouTube MobiuSRIT channel, 25 August 2007.

Why I YouTube. YouTube Andsanp channel, 28 November 2007.

Winfrey, Oprah. *Oprah's Message to YouTube*. YouTube Oprah channel, 1 November 2007.

Wu, Kevin. *The Real Me*. YouTube kevJumba channel, 14 May 2007.

XXX PORN XXX. YouTube AbolishTheSenateOrg channel, 19 October 2007.

XXX PORN XXX. YouTube UtubeAllFriends channel, 15 December 2007.

yay it's all GOOD. YouTube iwantfans channel, 5 April 2007.

Yonda, Aaron, and Matt Sloan. *Chocolate Rain by Chad Vader*. YouTube blamesocietyfilms channel, 14 August 2007.

You Have No Idea Who I Am. YouTube italktosnakes channel, 25 March 2008.

You Suck (video response). YouTube grownup channel, 10 November 2008.

YOUTUBE HATERS … !! YouTube ThePeacefulCat channel, 29 May 2008.

Youtube is Dead. YouTube Thunerf00t channel, 4 December 2008.

Bibliography

Abrams, Dominic. 'Social Identity, Social Cognition, and the Self: The Flexibility and Stability of Self-categorization.' In Dominic Abrams and Michael A. Hogg, eds, *Social Identity and Social Cognition*, 197–229. London: Blackwell, 1999.

Adorno, Theodor. 'Culture Industry Reconsidered.' Trans. Anson G. Rabinbach. *New German Critique* 6 (Fall 1975): 12–19.

Adorno, Theodor, and Max Horkheimer. *The Dialectic of Enlightenment.* London: Verso, 1979.

Agre, Philip E. 'Cyberspace as American Culture.' *Science as Culture* 11 (June 2002): 171–89.

Akass, Kim, and Janet McCabe. *Reading Sex and the City.* London: I.B. Tauris, 2004.

Albanesius, Chloe. 'Online Video No Threat to TV.' *PC Magazine*, 10 May 2007.

Aleo-Carreira, Cyndy. 'Bogus Anti-Scientology DMCA Notices Sent to YouTube Linked to Wikipedia User.' *Industry Standard*, 9 August 2008.

Allison, Juliann Emmons. *Technology, Development, and Democracy: International Conflict and Cooperation in the Information Age.* New York: SUNY Press, 2002.

Allport, Gordon. *The Use of Personal Documents in Psychological Science.* New York: Edwards, 1951.

'Americans Viewed a Record 16.8 Billion Videos Online in April Driven Largely by Surge in Viewership at YouTube.' comScore, 4 June 2009.

Anderson, Benedict. *Imagined Communities: Reflections on the Origin and Spread of Nationalism.* London: Verso, 1983.

Ang, Ien. *Living Room Wars: Rethinking Media Audiences for a Postmodern World.* London: Routledge, 1996.

– *Watching Dallas: Soap Opera and the Melodramatic Imagination.* Trans. Della Couling. London: Methuen, 1985.

'Angry YouTube Divorcee Speaks Out.' Transcript. On the Record with Greta Van Susteren, FoxNews. com, 28 May 2008.

Aqua. *Barbie Girl*. MCA Records, 1997.

Aries, Elizabeth. *Men and Women in Interaction: Reconsidering the Differences*. Oxford: Oxford University Press, 1996.

Arrington, Michael. 'YouTube Video Streams Top 1 Billion/Day.' *Washington Post*, 9 June 2009.

Arthur, Paul. 'The Moving Picture Cure: Self-Therapy Documentaries.' *Psychoanalytic Review* 94 (2007): 865–85.

Ashcroft, Bill, and Gareth Griffins. *The Post-Colonial Studies Reader*. London: Routledge, 1995.

Associated Press. 'Egyptian Student Gets 15 Years in US Terror Case.' 18 December 2008.

Atton, Chris. *An Alternative Internet: Radical Media, Politics and Creativity*. Edinburgh: Edinburgh University Press, 2004.

Attwood, Feona. 'Sexed Up: Theorizing the Sexualization of Culture.' *Sexualities* 9 (2006): 77–94.

Aufderheide, Patricia, and Peter Jaszi. 'Recut, Reframe, Recycle: Quoting Copyrighted Material in User-generated Video.' Center for Social Media's Future of Public Media Project, School of Communication, American University, Washington, DC, January 2008.

Aufderheide, Patricia, Peter Jaszi, and Elizabeth Nolan Brown. 'The Good, the Bad, and the Confusing: User-Generated Video Creators on Copyright.' Center for Social Media's Future of Public Media Project, School of Communication, American University, Washington, DC, April 2007.

Bailey, Carol A. *A Guide to Qualitative Field Research*. Thousand Oaks, CA: Pine Forge Press, 2007.

Bakardjieva, Maria. 'Virtual Togetherness: An Everyday-life Perspective.' *Media, Culture & Society* 25 (March 2003): 291–313.

Baker, Luke. 'Israel Gains in Media Blitz, But for How Long?' Reuters, 13 January 2009.

Bakker, William. 'More about Tourism BC's Online Video Strategy: Field Reporters.' Wilhelmus Blog, 26 October 2008.

Barlow, John Perry. 'A Declaration of Independence of Cyberspace.' Electronic Frontier Foundation, 8 February 1996.

Barnes, Brooks. 'For a Thrifty Audience, Buying DVDs Is So 2004.' *New York Times*, 22 November 2008.

Battelle, John. 'A Brief Interview with Michael Wesch (The Creator of That Wonderful Video …).' John Battelle's Search Blog, 18 February 2007.

Baumeister, Roy F., and Edward E. Jones. 'When Self-Presentation is Constrained by the Target's Knowledge: Consistency and Compensation.' *Journal of Personality and Social Psychology* 36 (1978): 608–18.

Baym, Nancy K. 'The New Shape of Online Community: The Example of Swedish Independent Music Fandom.' *First Monday* 12 (August 2007).

Benevenuto, Fabricio, Fernando Duarte, Tiago Rodrigues, Virgilio Almeida, Jussara Almeida, and Keith Ross. 'Characterizing Video Responses in Social Networks.' ArXiv.org, Cornell University Library, 30 April 2008.

Benevenuto, Fabricio, Tiago Rodrigues, Virgilio Almeida, Jussara Almeida, Chao Zhang, and Keith Ross. 'Identifying Video Spammers in Online Social Networks.' In *Proceedings of the 4th International Workshop on Adversarial Information Retrieval on the Web*, Beijing, China, 22 April 2008.

Benjamin, Walter. 'The Work of Art in the Age of Mechanical Reproduction.' In Meenakshi Gigi Durham and Douglas Kellner, eds, *Media and Cultural Studies: Keyworks*. 48–70. London: Blackwell, 2001.

Berlet, Chip. 'When Hate Went Online.' PublicEye.org, 10 October 2000.

Berman, John, Ethan Nelson, and Karson Yiu. 'The Blasphemy Challenge.' ABC News, 30 January 2007.

Beschizza, Rob. 'Creationist vs. Atheist YouTube War Marks New Breed of Copyright Claim.' *Wired*, 25 September 2007.

Biressi, Anita, and Heather Nun. *Reality TV: Realism and Revelation*. London: Wallflower Press, 2005.

Blanchard, Anita L. 'Developing a Sense of Virtual Community Measure.' *CyberPsychology & Behavior* 10 (2007): 827–30.

Blanton, Dana. 'Poll: For a Few True Believers, Elvis Lives.' Fox News, 14 August 2002.

Block, Bruce A. *The Visual Story: Seeing the Structure of Film, TV, and New Media*. New York: Focal Press, 2001.

Bolter, Jay David. 'Digital Essentialism and the Mediation of the Real.' In Heidi Philipsen and Lars Qvortrup, eds, *Moving Media Studies: Remediation Revisited*, 195–210. London: Gazelle Distribution Trade, 2007.

Bolter, Jay David, and Richard Grusin. *Remediation: Understanding New Media*. Cambridge, MA: MIT Press, 2000.

Bond, Paul. 'Online Video Chips Away at TV.' *Hollywood Reporter*, 17 November 2008.

– 'Study: Young People Watch Less TV.' *Hollywood Reporter*, 17 December 2008.

Bordo, Susan. *Unbearable Weight: Feminism, Western Culture, and the Body*. Berkeley: University of California, 1993.

Bourdieu, Pierre. *Distinction: A Social Critique of the Judgment of Taste*. Trans. Richard Nice. Boston: Harvard University Press, 1984.

– 'Understanding.' *Theory, Culture and Society* 13, 2 (1996): 17–37.

Brouwers, Janneke. 'YouTube vs. O-Tube: Negotiating a YouTube Identity.' *Cultures of Arts, Science and Technology* 1 (May 2008): 106–20.

Brown, Jane D., Carolyn Tucker Halpern, and Kelly Ladin L'Engle. 'Mass Media as a Sexual Super Peer for Early Maturing Girls.' *Journal of Adolescent Health* 36 (May 2005): 420–7.

Brown, Jeffrey C. 'Copyright Infringement Liability for Video Sharing Networks: Grokster Redux or Breaking New Ground under the Digital Millennium Copyright Act.' *The Computer and Internet Lawyer* 23 (December 2006): 10–17.

Brown, Lyn Mikel, and Mark B. Tappan. 'Fighting Like a Girl Fighting Like a Guy: Gender Identity, Ideology, and Girls at Early Adolescence.' *New Directions for Child and Adolescent Development* 120 (Summer 2008): 47–59.

Bruss, Elizabeth W. 'Eye for I: Making and Unmaking Autobiography in Film.' In James Olney, ed., *Autobiography: Essays Theoretical and Critical.* 296–320. Princeton, NJ: Princeton University Press, 1980.

Burgess, Jean. '"Is Your Chocolate Rain Are Belong to Us"?: Viral Video, YouTube and the Dynamics of Participatory Culture.' In Geert Lovink and Sabine Niederer, eds, *Video Vortex Reader: Responses to YouTube.* 101–9. Amsterdam: Institute of Network Cultures, 2008.

Burgess, Jean, and Joshua Green. 'Agency and Controversy in the YouTube Community.' Paper presented at Internet Research 9.0: Rethinking Community, Rethinking Place, University of Copenhagen, Denmark, 15 October 2008.

– *YouTube: Online Video and Participatory Culture.* Cambridge, UK: Polity Press, 2009.

Butters, Andrew Lee. 'Fighting the Media War in Gaza.' *Time*, 14 January 2009.

Calavita, Marco. 'A Futile Attempt at Medium Specificity: Hollywood's Cinematic Assault on Television and Video, 1994–2003.' *Atlantic Journal of Communication* 13 (August 2005): 135–49.

Campbell, Jan, and Janet Harbord. Introduction. Jan Campbell and Janet Harbord, eds, *Temporalities, Autobiography and Everyday Life: Autobiography and Everyday Life.* 1–20. Manchester: Manchester University Press, 2002.

Carli, Linda L., and Danuta Bukatko. 'Gender, Communication, and Social Influence.' In Thomas Eckes and Hanns M. Trautner, eds, *The Developmental Social Psychology of Gender.* 295–332. New York: Lawrence Erlbaum Associates, 2000.

Carlson, Tom, and Kim Strandberg. 'Riding the Web 2.0 Wave: Candidates on YouTube in the 2007 Finnish National Elections.' Paper presented at the 4th General Conference of the European Consortium of Political Research, Pisa, Italy, 6–8 September 2007.

Carpenter, Edmund. 'The New Languages.' In Carpenter and Marshall McLuhan, eds, *Explorations in Communication*. 162–79. Boston: Beacon Press, 1968.

Carruthers, Susan L. 'No One's Looking: The Disappearing Audience for War.' *Media, War & Conflict* 1 (January 2008): 70–6.

Carter, Scott, and Jennifer Mankoff. 'When Participants Do the Capturing: The Role of Media in Diary Studies.' Paper presented to the Conference on Human Factors in Computing Systems, Portland, OR, 2–7 April 2005.

Castells, Manuel. 'Communication, Power and Counter-Power in the Network Society.' *International Journal of Communication* 1 (2007): 238–66.

Cha, Meeyoung, Haewoon Kwak, Pablo Rodriguez, Yong-Yeol Ahn, and Sue Moon. 'I Tube, You Tube, Everybody Tubes: Analyzing the World's Largest User Generated Content Video System.' In *Proceedings of the 7th ACM SIGCOMM Conference on Internet Measurement*, San Diego, CA, 24–26 October 2007.

Chalfen, Richard. 'Home Movies as Cultural Documents.' In Sari Thomas, ed., *Film Culture: Explorations of Cinema in its Social Context*. 126–38. London: Scarecrow Press, 1982.

– *Snapshot Versions of Life*. Bowling Green, OH: Bowling Green State University Press, 1987.

Chen, Hsinchun, Sven Thoms, and T.J. Fu. 'Cyber Extremism in Web 2.0: An Exploratory Study of International Jihadist Groups.' Paper presented to the IEEE International Conference on Intelligence and Security Informatics, Taiwan, 17–20 June 2008.

Cheng, Jacqui. 'Nielsen: YouTube is from Mars, Streaming Video is from Venus.' ArsTechnica.com, 14 February 2008.

Cheng, Xu, Cameron Dale, and Jiangchuan Liu. 'Understanding the Characteristics of Internet Short Video Sharing: YouTube as a Case Study.' arXiv.org, Cornell University Library, 25 July 2007.

Chin, Bertha, and Matt Hills. 'Restricted Confessions? Blogging, Subcultural Celebrity and the Management of Producer-fan Proximity.' *Social Semiotics* 18 (June 2008): 253–72.

Chottiner, Lee. 'The YouTube War.' *Jewish Chronicle*, 13 January 2009.

Christensen, Christian. 'Uploading Dissonance: YouTube and the US Occupation of Iraq.' *Media, War & Conflict* 1 (2008): 155–75.

– 'YouTube: The Evolution of Media?' *Screen* 45 (2007): 36–40.

'Cisco Visual Networking Index: Forecast and Methodology, 2007–2012.' Cisco Systems, 16 June 2009.

Cleverley, Kaarina. 'Ideology and Visual Anthropology: 18th Nordic Anthropological Film Association Conference and Festival.' *Visual Anthropology Review* 13 (Spring 1997): 72–4.

Clifford, James, and George E. Marcus, eds, *Writing Culture: The Poetics and Politics of Ethnography*. Berkeley: University of California Press, 1986.

Cohen, Leonard. *There Is a War*. New Skin for the Old Ceremony. Sony Music Entertainment, 1974.

Colombani, Jean-Marie. 'We Are All Americans.' *Le Monde*, 12 September 2001.

Comstock, George, and Erica Scharrer. *Media and the American Child*. London: Academic Press, 2007.

Cooper, Charlotte. 'What's Fat Activism?' Working Paper WP2008-02, Department of Sociology Working Paper Series, University of Limerick, September 2008.

Corner, John. *The Art of Record: A Critical Introduction to Documentary*. Manchester: Manchester University Press, 1996.

Couldry, Nick. 'Liveness, "Reality," and the Mediated Habitus from Television to the Mobile Phone.' *Communication Review* 7 (October 2004): 353–61.

– 'Speaking about Others and Speaking Personally: Reflections after Elspeth Probyn's Sexing the Self.' *Cultural Studies* 10 (May 1996): 315–33.

Cox, Ana Marie. 'The YouTube War.' *Time*, 19 July 2006.

Cubitt, Sean. 'Codecs and Capability.' In Geert Lovink and Sabine Niederer, eds, *Video Vortex Reader: Responses to YouTube*. 45–51. Amsterdam: Institute of Network Cultures, 2008.

– *Timeshift: On Video Culture*. London: Routledge, 1991.

– *Videography: Video Media as Art and Culture*. New York: St. Martin's Press, 1993.

Curran, James. *Media and Power*. London: Routledge, 2002.

Curran, James, David Morley, and Valerie Walkerdine. *Cultural Studies and Communication*. Oxford: Oxford University Press, 1998.

Czach, Liz. '*There's No Place Like Home Video* (Review).' *The Moving Image* 3 (Fall 2003): 114–16.

Damast, Alison. 'The Admissions Office Finds Facebook.' *BusinessWeek*, 28 September 2008.

Dames, K. Matthew. 'Copyright and the Speed Limit.' *Information Today*, February 2008, 17–18.

Darby, Seyward. 'Technology: Sex, Lies and YouTube.' *Transitions Online*, 31 July 2007.

'Dark Web Terrorism Research.' AI Lab Dark Web Project, University of Arizona, 2009.

Daswani, Mansha. 'Deloitte Issues 2009 Media Predictions.' *World Screen*, 23 January 2009.

Davis, Douglas. *The Five Myths of Television Power and Why the Medium Is Not the Message*. New York: Simon and Schuster, 1993.

Dawson, Lorne L., and Jenna Hennebry. 'New Religions and the Internet:

Recruiting in a New Public Sphere.' In Lorne L. Dawson, ed., *Cults and New Religious Movements: A Reader.* 217–91. London: Blackwell, 2003.

de Klerk, Nico. 'Home Away from Home: Private Films from the Dutch East Indies.' In Karen L. Ishizuka and Patricia R. Zimmerman, eds, *Mining the Home Movie: Excavations in Histories and Memories.* 148–62. Berkeley: University of California Press, 2008.

de Kuyper, Eric. 'Aux origines du cinema: le film de famille.' In Roger Odin, ed.,*Le Film de famille: usage privé, usage public.* 11–26. Paris: Méridiens-Klincksieck, 1995.

De Vos, Gail. *Storytelling for Young Adults: A Guide to Tales for Teens.* Westport, CT: Libraries Unlimited, 2003.

Diasi, Karen. 'The Ana Sanctuary: Women's Pro-Anorexia Narratives in Cyberspace.' *Journal of International Women's Studies* 4 (April 2003).

Dickinson, Tim. 'The First YouTube Election: George Allen and "Macaca."' *Rolling Stone*, 15 August 2006.

Di Leonardo, Micaela. *Gender at the Crossroads of Knowledge: Feminist Anthropology in the Postmodern Era.* Berkeley: University of California Press, 1991.

Douglas, Mary. *Risk and Blame: Essays in Cultural Theory.* London: Routledge, 1992.

Dreyfus, Hubert L., and Paul Rabinow. *Michel Foucault: Beyond Structuralism and Hermeneutics.* 2nd ed. Chicago: University of Chicago Press, 1983.

Driskell, Robyn Bateman, and Larry Lyon, 'Are Virtual Communities True Communities? Examining the Environments and Elements of Community.' *City & Community* 1 (December 2002): 373–90.

Eaton, Phoebe. 'The YouTube Divorcée.' *New York Magazine*, 1 June 2008.

Egan, Susanna. 'Encounters in Camera: Autobiography as Interaction.' *Modern Fiction Studies* 40 (1994): 593–618.

Elliot, Stuart. 'The Media Business.' *New York Times*, 19 October 1994.

Evans, Mary. *Missing Persons: The Impossibility of Auto/biography.* London: Routledge, 1999.

Evolution of Jihad Video. IntelCenter. Alexandria, VA: Tempest, 2005.

Fairbanks, Eve. 'The YouTube Election: Candid Camera.' *New Republic Online*, 2 November 2006.

Faris, Robert, Stephanie Wang, and John Palfrey, 'Censorship 2.0.' *Innovations* 3 (June 2008): 165–87.

Faulkner, Susan, and Jay Melican. 'Getting Noticed, Showing-Off, Being Overheard: Amateurs, Authors and Artists Inventing and Reinventing Themselves in Online Communities.' In *Ethnographic Praxis in Industry Conference Proceedings*, Colorado, 2007.

Fernback, Jan. 'The Individual within the Collective: Virtual Ideology and the

Realization of Collective Principles.' In Steven Jones, ed., *Virtual Culture: Identity and Communication in Cyberspace.* 36–54. London: Sage, 1997.

Ferreday, Debra. 'Unspeakable Bodies.' *International Journal of Cultural Studies* 6 (2003): 277–95.

Feuer, Alan, and Jason George. 'Internet Fame Is Cruel Mistress for a Dancer of the Numa Numa.' *New York Times*, 26 February 2005.

Fiske, John. *Power Plays, Power Works.* London: Verso, 1993.

– *Television Culture.* London: Routledge, 1987.

– *Understanding Popular Culture.* London: Unwin Hyman, 1989.

Flint, Joe, and Kate Kelly. 'New Signs of Strain for Blockbuster.' *Wall Street Journal*, 19 September 2005, B5.

Flitterman-Lewis, Sandy. 'Psychoanalysis, Film, and Television.' In Robert C. Allen, ed., *Channels of Discourse: Television and Contemporary Criticism.* 172–210. London: Methuen, 1987.

Forster, Peter M. 'Psychological Sense of Community in Groups on the Internet.' *Behaviour Change* 21 (2004): 141–6.

Foucault, Michel. *The History of Sexuality.* Vol. 1: *An Introduction.* Trans. Robert Hurley. New York: Vintage Books, 1978.

Fox, Susannah. 'Older Americans and the Internet.' Pew Internet and American Life Project, Pew Research Center, Washington, DC, 28 March 2004.

Frantz, Ashley. 'Teen Alleging Rape Turns to YouTube.' CNN.com, 16 May 2008.

Freilich, Leon. Letter to the Editor. *New York Times*, 20 August 2006.

Friedrich, William N., Jennifer Fisher, Daniel Broughton, Margaret Houston, and Constance R. Shafran. 'Normative Sexual Behavior in Children: A Contemporary Sample.' *Pediatrics* 101 (April 1998): e9.

Galbi, Douglas. 'Stories Largely Missing in Online Video.' Purple Motes Blog, purplemotes.net, 1 June 2008.

Gannes, Liz. 'Online Video: How Big Is It Really?' *BusinessWeek*, 4 June 2009.

– 'Video Viewers Up 27% from 2007; as for the Money ...' *Fortune*, 17 November 2008.

Gantz, John F. 'The Diverse and Exploding Digital Universe: An Updated Forecast of Worldwide Information Growth Through 2011.' IDC, Framingham, MA, March 2008.

Gare, Arran. 'The Politics of Recognition versus the Politics of Hatred.' *Democracy and Nature* 8 (July 2002): 261–80.

Garroutte, Eva Marie. *Real Indians: Identity and the Survival of Native America.* Berkeley: University of California Press, 2003.

Geertz, Clifford. *The Interpretation of Cultures.* New York: Basic Books, 1973.

– *Local Knowledge: Further Essays in Interpretive Anthropology.* New York: Basic Books, 1983.

'Gender Ratio of Netizens Becomes More Balanced.' WomenofChina.cn, 30 July 2008.

George, Carlisle, and Jackie Scerri. 'Web 2.0 and User-Generated Content: Legal Challenges in the New Frontier.' *Journal of Information, Law and Technology* 2 (2007): 1–22.

'German Jewish Group Takes YouTube to Court.' *Spiegel Online*, 21 March 2008.

Gibson, Barbara Ellen. 'Co-producing Video Diaries: The Presence of the "Absent" Researcher.' *International Journal of Qualitative Methods* 4 (December 2005): 1–9.

Gilder, George. *Life After Television: The Coming Transformation of Media and American Life.* Rev. ed. London: W.W. Norton, 1994.

Gillespie, Tarleton. *Wired Shut: Copyright and the Shape of Digital Culture.* Cambridge, MA: MIT Press, 2007.

Ginsberg, Benjamin. *The Captive Public: How Mass Opinion Promotes State Power.* New York: Basic Books, 1984.

Gissin, Amir R. 'No Israeli Apology for Defeating Iran's Islamic Proxy.' *Toronto Star*, 24 January 2009.

Goldhaber, Michael H. 'The Attention Economy and the Net.' *First Monday* 2 (April 1997).

Gomes, Lee. 'Portals: YouTube Is Now a Verb and an Adjective.' *Wall Street Journal*, 18 October 2006.

Gonsalves, Antone. 'Young TV Viewers Prefer Web over DVRs.' *InformationWeek*, 23 January 2009.

Goswami, Usha. 'A Research Literature Review: Child Development,' Annex H, *Byron Review on the Impact of New Technologies on Children*, Centre for Neuroscience in Education, University of Cambridge, January 2008.

Griffith, Margaret. 'Looking for You: An Analysis of Video Blogs.' Paper presented to the annual meeting of the Association for Education in Journalism and Mass Communication, Washington, DC, 8 August 2007.

Gross, Ralph, and Alessandro Acquisti. 'Information Revelation and Privacy in Online Social Networks (The Facebook Case).' Paper presented to the ACM Workshop on Privacy in the Electronic Society, Alexandria, VA, November 2005.

Grossman, Lev. 'Time's Person of the Year: You.' *Time*, 13 December 2006.

Grosz, Elizabeth. *Volatile Bodies: Toward a Corporeal Feminism.* Bloomington: Indiana University Press 1994.

Gueorguieva, Vassia. 'Voters, MySpace, and YouTube.' *Social Science Computer Review* 26 (August 2008): 288–300.

Gumbel, Andrew. 'The "YouTube Elections": How Campaigns Are Being Scrutinised as Never Before.' *Independent*, 4 November 2006.

Gunning, Tom. 'The Cinema of Attractions: Early Film, Its Spectator and the Avant-Garde.' In Thomas Elsaesser, ed., *Early Cinema: Space Frame Narrative.* 56–62. London: British Film Institute, 1990.

Gunster, Shane. *Capitalizing on Culture: Critical Theory for Cultural Studies.* Toronto: University of Toronto Press, 2004.

Hampton, Keith, and Barry Wellman. 'Neighboring in Netville: How the Internet Supports Community and Social Capital in a Wired Suburb.' *City & Community* 2 (November 2003): 277–311.

Hardy, Quentin, and Evan Hessel. 'GooTube.' *Forbes,* 16 June 2008.

Harley, Dave, and Geraldine Fitzpatrick. 'YouTube and Intergenerational Communication: The Case of Geriatric1927.' *Universal Access in the Information Society,* 29 May 2008.

Hart, Kim. 'On YouTube, Lawmakers Have Sites to Behold.' *Washington Post,* 13 January 2009.

Hartley, John. 'YouTube, Digital Literacy and the Growth of Knowledge.' Paper presented to the Media@Lse Fifth Anniversary Conference, London, September 2008.

Hausenblaus, Michael, and Frank Nack. 'Interactivity = Reflective Expressiveness.' *IEEE MultiMedia* 14 (April 2007): 4–7.

Hauser, Gerard A. *Vernacular Voices: The Rhetoric of Publics and Public Spheres.* Columbia: University of South Carolina Press, 1999.

Heffernan, Virginia. 'The Hitler Meme.' *New York Times,* 24 October 2008.

– 'The Many Tribes of YouTube.' *New York Times,* 27 May 2007.

– 'Narrow Minded.' *New York Times,* 25 May 2008.

Heffernan, Virginia and Tom Zeller Jr, 'The Lonelygirl That Really Wasn't.' *New York Times,* 13 September 2006.

Hess, Mike. 'Video Consumer Mapping Study.' Center for Media Design, Bell State University. 26 March 2009.

Hilderbrand, Lucas. 'Youtube: Where Cultural Memory and Copyright Converge.' *Film Quarterly* 16 (Fall 2007): 48–57.

Hill, Matt. *Fan Cultures.* London: Routledge, 2002.

Hobsbawm, Eric J. *The Age of Extremes: The Short Twentieth Century.* London: Abacus, 1994.

Holliday, Ruth. 'We've Been Framed: Visualising Methodology,' *Sociological Review* 48 (2001): 503–21.

Holson, Laura M. 'Hollywood Asks YouTube: Friend or Foe?' *New York Times,* 15 January 2007.

Homiak, John, and Pamela Wintle. 'The Human Studies Film Archives, Smithsonian Institution.' In Karen L. Ishizuka and Patricia R. Zimmerman, eds, *Mining the Home Movie: Excavations in Histories and Memories.* 41–6. Berkeley: University of California Press, 2008.

hooks, bell. *Ain't I a Woman: Black Women and Feminism.* New York: South End Press, 2007.

Hornor, Gail. 'Sexual Behavior in Children. Normal or Not?' *Journal of Pediatric Health Care* 18 (2004): 57–64.

Horrigan, John B. 'Home Broadband Adoption 2008.' Pew Internet & American Life Project, Pew Research Center, Washington, DC, July 2008.

Hráček, Filip. 'Audiovizuální styl uživatelských YouTube videí.' Ph.D. diss., Masarykova University, 2009.

Huffaker, David A., and Sandra L. Calvert. 'Gender, Identity, and Language Use in Teenage Blogs.' *Journal of Computer-Mediated Communication* 10 (June 2006).

Hunt, Kurt. 'Copyright and YouTube: Pirate's Playground or Fair Use Forum.' *Michigan Telecommunications and Technology Law Review* 14 (2007): 197–222.

Hunt, Lynn. 'Introduction: Obscenity and the Origins of Modernity, 1500–1800.' In Lynn Hunt, ed., *The Invention of Pornography: Obscenity and the Origins of Modernity, 1500–1800.* 9–45. New York: Zone Books, 1996.

Innis, Harold A. *The Bias of Communication.* Toronto: University of Toronto Press, 1951.

'Israel Clashes With YouTube Over Censorship.' Fox News, 2 January 2009.

Itzkoff, Dave. 'A One-Way Ticket to Disaster.' *New York Times,* 30 December 2007.

Jacobs, Sue-Ellen, Wesley Thomas, and Sabine Lang. *Two-spirit People: Native American Gender Identity, Sexuality, and Spirituality.* Champaign: University of Illinois Press, 1997.

Jacobson, Bob. 'Reply to "One-off Moments in Time."' Bambi Francisco Blog, 15 June 2006.

Jancovich, Mark and Lucy Faire. *The Place of the Audience: Cultural Geographies of Film Consumption.* London: British Film Institute, 2003.

Jenkins, Henry. *Convergence Culture: Where Old and New Media Collide.* New York: New York University Press, 2006.

– 'From YouTube to WeTube …' Confessions of an Aca-Fan, HenryJenkins.org, 13 February 2008.

– 'Nine Propositions Towards a Cultural Theory.' Confessions of an Aca-Fan, HenryJenkins.org, 28 May 2007.

Jones, K.C. 'Videotaped Florida Teen Beating Prompts Calls To Block Violent Content.' *Information Week,* 8 April 2008.

Jones, Steve. 'The Internet and its Social Landscape.' In Steve Jones, ed., *Virtual Culture: Identity and Communication in Cybersociety.* 7–35. London: Sage, 1997.

Jones, Sydney. 'Anthropology of YouTube.' PewInternet.org, Pew Internet Posts, 25 July 2008.

Juffer, Jane. *At Home with Pornography: Women, Sex, and Everyday Life.* New York: New York University Press, 1998.

Juhasz, Alexandra. 'Learning From Fred.' *Teachers College Record*, 8 September 2008.

– 'Why Not (to) Teach on YouTube.' In Geert Lovink and Sabine Niederer, eds, *Video Vortex Reader: Responses to YouTube*. 133–9. Amsterdam: Institute of Network Cultures, 2008.

Junee, Ryan. 'Zoinks! 20 Hours of Video Uploaded Every Minute!' YouTube Blog, 20 May 2009.

Kakutani, Michiko. 'The Cult of the Amateur.' *New York Times*, 29 June 2007.

Kambouri, Nelli, and Pavlos Hatzopoulos. 'Making Violent Practices Public.' In Geert Lovink and Sabine Niederer, eds, *Video Vortex Reader: Responses to YouTube*. 125–31. Amsterdam: Institute of Network Cultures, 2008.

Kampman, Minke. 'Flagging or Fagging: (Self-)Censorship of Gay Content on YouTube.' In Geert Lovink and Sabine Niederer, eds, *Video Vortex Reader: Responses to YouTube*. 153–60. Amsterdam: Institute of Network Cultures, 2008.

Kaplinsky-Dwarika, Roger. '"Most Annoying Sound" Is Stomach Turning.' *ABCNews*, 25 January 2007.

Karpovich, Angelina I. 'Reframing Fan Videos.' In Jamie Sexton, ed., *Music, Sound and Multimedia: From the Live to the Virtual*. 17–28. Edinburgh: Edinburgh University Press, 2007.

Kattelle, Allan D. *Home Movies: A History of the American Industry, 1897–1979*. Nashua, NH: Transition, 2000.

Keating, Gina. 'U.S. Woes Force Broadcasters to Look to New Media.' Reuters, 2 February 2009.

Keen, Andrew. *The Cult of the Amateur: How Today's Internet is Killing Our Culture*. Garden City, NY: Doubleday, 2007.

Kelly, Kevin. 'We Are the Web.' *Wired* 13 (August 2008).

Kellner, Douglas. *Media Culture: Cultural Studies, Identity and Politics Between the Modern and the Postmodern*. London: Routledge, 1995.

Kennedy, Tracy L.M. 'An Exploratory Study of Feminist Experiences In Cyberspace.' *CyberPsychology & Behavior* 3 (October 2000): 707–19.

Kiley, David. 'The YouTube Election Ratifies Google's Investment.' *BusinessWeek*, 8 November 2006.

Klinger, Barbara. *Beyond the Multiplex: Cinema, New Technologies, and the Home*. Berkeley: University of California Press, 2006.

Kruitbosch, Gijs, and Frank Nack. 'Broadcast Yourself on YouTube: Really?' In *Proceedings of the 3rd ACM International Workshop on Human-centered Computing*, Vancouver, BC, 31 October 2008.

Kumar, Parul. 'Locating the Boundary between Fair Use and Copyright Infringement: The Viacom–YouTube Dispute.' *Journal of Intellectual Property Law & Practice* 184 (October 2008).

Lamerichs, Nicolle. 'It's a Small World After All: Metafictional Fan Videos on YouTube.' *Cultures of Arts, Science and Technology* 1 (May 2008): 52–71.

Landow, George P. 'Autobiography, Autobiographicality, and Self-Representation.' Working Paper, The Victorian Web Blog, n.d.

Landry, Brian M., and Mark Guzdial. 'Art or Circus? Characterizing User-Created Video on YouTube.' SIC Technical Reports, Georgia Institute of Technology, 2008.

Lange, Patricia G. 'Commenting on Comments: Investigating Responses to Antagonism on YouTube.' Paper presented to the Society for Applied Anthropology Conference, Tampa, FL, 31 March 2007.

– '(Mis)conceptions About YouTube.' In Geert Lovink and Sabine Niederer, eds, *Video Vortex: Responses to YouTube*, 87–99. Amsterdam: Institute of Network Cultures, 2008.

– 'Publicly Private and Privately Public: Social Networking on YouTube.' *Journal of Computer-Mediated Communication* 13 (December 2007): 361–80.

Lastowka, Greg, and Candidus Dougherty. 'Copyright: Copyright Issues in Virtual Economies.' *E-Commerce & Policy* 9 (May 2007): 28–40.

Learmonth, Michael. 'Nielsen: Primetime Is At Work For Web Video.' *Silicon Alley Insider*, 3 March 2008.

– 'YouTube Moving the Needle on Ad Sales.' *Advertising Age*, 8 April 2009.

Le Beau, Bryan F. *The Atheist: Madalyn Murray O'Hair*. New York: New York University Press, 2005.

Lebo, Harlan. 'First Release of Findings From the UCLA World Internet Project Shows Significant "Digital Gender Gap" in Many Countries.' UCLA Newsroom, 14 January 2004.

Lee, Jin. 'Locality Aware Peer Assisted Delivery: The Way to Scale Internet Video to the World.' Paper presented at Packet Video 2007, Lausanne, Switzerland, 12–13 November 2007.

Lefebvre, Henri. *Critique of Everyday Life*. Vol. 1. London: Verso, 1991.

Leighton, Tom. 'Improving Performance on the Internet.' *ACM Queue*, October 2008.

Leiss, William, Stephen Kline, Sut Jhally, and Jacqueline Botterill. *Social Communication in Advertising: Consumption in the Mediated Marketplace*. London: Routledge, 2005.

Lenert, Edward M. 'A Communication Theory Perspective on Telecommunications Policy.' *Journal of Communications* 48 (7 February 2006): 3–23.

Lenhart, Amanda Mary Madden, Alexandra Rankin Macgill, and Aaron Smith. 'Teens and Social Media.' Pew Internet & American Life Project, Pew Research Center, Washington, DC, 19 December 2007.

Lessig, Lawrence. 'In Defense of Piracy.' *Wall Street Journal*, 11 October 2008.

– *Remix: Making Art and Commerce Thrive In the Hybrid Economy.* New York: Penguin, 2008.

Levin, G. Roy. *Jean Rouch, Documentary Explorations: 15 Interviews with Film-Makers.* Garden City, NY: Doubleday, 1971.

Lin, Wei-Hao, and Alexander Hauptmann. 'Identifying Ideological Perspectives of Web Videos Using Folksonomies.' Association for the Advancement of Artificial Intelligence Fall Symposium Series Papers, Menlo Park, CA, 2008.

Livingstone, Sonia. *Young People and the New Media: Childhood and the Changing Media Environment.* London: Sage, 2002.

Livingstone, Sonia, and Nancy Thumim. 'What is Fred Telling Us?' *Teachers College Record*, 8 September 2008.

Lizza, Ryan. 'The YouTube Election.' *New York Times*, 20 August 2006.

Losh, Elizabeth. 'Government YouTube: Bureaucracy, Surveillance, and Legalism in State-Sanctioned Online Video Channels.' In Geert Lovink and Sabine Niederer, eds, *Video Vortex Reader: Responses to YouTube.* 111–23. Amsterdam: Institute of Network Cultures, 2008.

Louderback, Jim. 'Where's the Outrage Over Online Video Viewership Claims?' *AdAge*, 28 October 2008.

Lovink, Geert. 'Radical Media Pragmatism Strategies for Techno-Social Movements.' *Like, Art Magazine* 6 (Winter/Spring 1998): 33–7.

Lunceford, Brett, and Shane Lunceford, 'Meh. The Irrelevance of Copyright in the Public Mind.' *Northwestern Journal of Technology and Intellectual Property* 7 (Fall 2008): 33–49.

MacDonald, G. Jeffrey. 'Teens: It's a Diary. Adults: It's Unsafe.' *Christian Science Monitor*, 25 May 2005.

MacDougall, David. 'Prospects of the Ethnographic Film.' *Film Quarterly* 23, no. 2 (1969): 16–30.

MacKinnon, Catharine A. 'Sexuality, Pornography, and Method: "Pleasure Under Patriarchy."' In Nancy Tuana and Rosemarie Tong, eds, *Feminism and Philosophy: Essential Readings in Theory, Reinterpretation, and Application.* 134–61. Boulder, CO: Westview Press, 1995.

Macnamara, Jim. 'E-Electioneering: Use of New Media in the 2007 Australian Federal Election.' Paper presented to the ANZCA08 Conference, Power and Place, Wellington, New Zealand, July 2008.

Madden, Mary. 'Online Video.' Pew Internet & American Life Project, Pew Research Center, Washington, DC, 25 July 2007.

Manovich, Lev. 'The Practice of Everyday (Media) Life.' In Geert Lovink and Sabine Niederer, eds, *Video Vortex Reader: Responses to YouTube*, 33–43. Amsterdam: Institute of Network Cultures, 2008.

Martin, Don. 'Don Martin's Stories of the Year.' *National Post*, 27 December 2008.

Martinson, Floyd Mansfield. *The Sexual Life of Children*. New York: Greenwood, 1994.

McArthur, Keith. 'YouTube No Place to Discuss Ideas.' Com.Motion Blog, 3 March 2008.

McChesney, Robert W. *The Problem of the Media: U.S. Communication Politics in the Twenty-First Century*. New York: Monthly Review Press, 2004.

McIntyre, Douglas A. Introduction. *Time*, 14 May 2009.

McKenna, Katelyn Y.A., Amie S. Green, and Marci E.J. Gleason. 'Relationship Formation on the Internet: What's the Big Attraction?' *Journal of Social Issues* 58 (December 2002): 9–31.

McMurria, John. 'The YouTube Community.' *Flow* 5 (20 October 2006).

Metz, Cade. 'YouTube Rolls Out Scientology Double Standard.' *Register*, 2 May 2008.

Meyrowitz, Joshua. *No Sense of Place: The Impact of Electronic Media on Social Behaviour*. Oxford: Oxford University Press, 1986.

Milliken, Mary, Kerri Gibson, Susan O'Donnell, and Janice Singer. 'User-Generated Online Video and the Atlantic Canadian Public Sphere: A YouTube Study.' In *Proceedings of the International Communication Association Annual Conference*, Montreal, 22–26 May 2008.

Mills, Simon. 'Sean Cubitt: Interview by Sean Mills.' *Framed* 10, (November 2006).

Minty, Riyaad. 'Waging the Web Wars.' AlJazeera.net, 21 January 2009.

Mohseni, Manouchehr, Behzad Dowran, and Mohammad Hadi Sohrabi Haghighat. 'Does the Internet Make People Socially Isolated? A Social Network Approach.' *Bangladesh e-Journal of Sociology* 5 (January 2008): 81–101.

Molyneaux, Heather, Susan O'Donnell, Kerri Gibson, and Janice Singer. 'Exploring the Gender Divide on YouTube: An Analysis of the Creation and Reception of Vlogs.' *American Communication Journal* 10 (Summer 2008).

Moon, Robert. 'Dance Dance Evolution.' Working Paper, Interactive Telecommunications Program, New York University, n.d.

Morahan, Martin J. 'Women and Girls Last: Females and the Internet.' Paper presented at the Internet Research and Information for Social Scientists International Conference, Bristol, England, March 1998.

Moran, James M. *There's No Place Like Home Video*. Minneapolis: University of Minnesota Press, 2002.

'More Than One Billion Users Will View Online Video in 2013.' Allied Business Intelligence, 27 May 2008.

Moscovitch, Philip. 'Hitler's Downfall, Parodied.' *Globe and Mail*, 22 December 2008.

Mulveen, Ruaidhri. 'An Interpretative Phenomenological Analysis of Participation in a Pro-anorexia Internet Site and Its Relationship with Disordered Eating.' *Journal of Health Psychology* 11 (2006): 283–96.

Mulvey, Laura. 'Visual Pleasure and Narrative Cinema.' *Screen* 6 (Autumn 1975): 6–18.

Murphy, Alexandra G. 'The Dialectical Gaze: Exploring the Subject-Object Tension in the Performances of Women Who Strip.' *Journal of Contemporary Ethnography* 32 (2003): 305–35.

Murthy, Dhiraj. 'Digital Ethnography: An Examination of the Use of New Technologies for Social Research.' *Sociology* 42 (October 2008): 837–55.

Naím, Moisés. 'The YouTube Effect.' *Foreign Policy* (January/February 2007).

Napoli, Philip M. 'Audience Measurement and Media Policy: Audience Economics, the Diversity Principle, and the Local People Meter.' Fordham University, McGannon Center Working Paper Series, 2008.

Nead, Lynda. *The Female Nude: Art, Obscenity and Sexuality*. London: Routledge, 1992.

Newman, Michael Z. 'Ze Frank and the Poetics of Web Video.' *First Monday* 13 (May 2008).

Nielsen. 'Americans Watching More TV Than Ever; Web and Mobile Video Up too.' Nielsen.com, 20 May 2009.

– 'TV Viewing Among Kids at an Eight-Year High.' Nielsen.com, 26 October 2009.

– 'The Video Generation: Kids and Teens Consuming More Online Video Content than Adults at Home.' Nielsen.com, 9 June 2008.

Noam, Eli M. 'The Economics of User Generated Content and Peer-to-Peer: The Commons as the Enabler of Commerce.' In Noam and Lorenzo Maria Pupillo, eds, *Peer-to-Peer Video: The Economic, Policy, and Culture of Today's New Mass Medium*. 3–13. New York: Springer, 2008.

Nussbaum, Felicity A. 'Toward Conceptualizing Diary.' In James Olney, ed., *Studies in Autobiography*. 128–40. Oxford: Oxford University Press, 1988.

O'Brien, Damian. 'Copyright Challenges for User Generated Intermediaries: Viacom v. YouTube and Google.' In Brian Fitzgerald, Fuping Gao, Damien O'Brien and Sampsung Xiaoxiang Shi, eds, *Copyright Law, Digital Content and the Internet in the Asia-Pacific*. 219–33. Sydney: Sydney University Press, 2008.

Oksanen, Ville, and Mikko Välimäki. 'Theory of Deterrence and Individual Behavior. Can Lawsuits Control File Sharing on the Internet?' *Review of Law & Economics* 3 (2007).

Ortega, Tony, 'What to Get L. Ron Hubbard for His Birthday: How "Anonymous" Has Changed the Game of Exposing Ron's Ruthless Global Scam.' *Village Voice*, 4 March 2008.

Ortutay, Barbara. 'Survey: Family Time Eroding as Internet Use Soars.' Yahoo Tech News, 15 June 2009.

Oulliette, Laurie. 'Camcorder Dos and Don'ts: Popular Discourses on Amateur Video and Participatory Television.' *Velvet Light Trap* 36 (September 1995).

Paolillo, John C. 'Structure and Network in the YouTube Core.' In *Proceedings of the 41st Annual Hawaii International Conference on System Sciences*, Waikoloa, Big Island, HI, 7 January 2008.

Peers, Martin. 'Internet Killed the Video Star.' *Wall Street Journal*, 5 February 2009, C10.

Pegoraro, Rob. 'TV Over the Web: Still a Fuzzy Picture.' *Washington Post*, 15 January 2009.

Peraica, Ana. 'Chauvinist and Elitist Obstacles Around YouTube and Porntube: A Case Study of Home-Made Porn Defended as "Video Art."' In Geert Lovink and Sabine Niederer, eds, *Video Vortex Reader: Responses to YouTube*. 189–93. Amsterdam: Institute of Network Cultures, 2008.

Pini, Maria. 'Girls on Film: Video Diaries as "Auto-ethnographies."' Working Paper, London Knowledge Lab, 8 May 2006.

Pink, Sarah. 'More Visualising, More Methodologies: On Video, Reflexivity and Qualitative Research.' *Sociological Review* 49 (June 2008): 586–99.

Plato. *The Republic: The Complete and Unabridged Jowett Translation*. London: Vintage, 1991.

Pluempavarn, Patchareeporn, and Niki Panteli. 'The Creation of Social Identity Through Weblogging.' University of Bath, School of Management Working Paper Series, October 2007.

Pogue, David. 'Shazam! A Projector Is Shrunk.' *New York Times*, 6 November 2008.

Poster, Mark. *Information Please: Culture and Politics in the Age of Digital Machines*. Durham, SC: Duke University Press, 2006.

Probyn, Elspeth. *Sexing the Self: Gendered Positions in Cultural Studies*. London: Routledge, 1993.

Quan-Haase, Anabel, and Barry Wellman. 'Capitalizing on the Net: Social Contact, Civic Engagement, and Sense of Community.' In Barry Wellman and Caroline Haythornthwaite, eds, *The Internet in Everyday Life*. 291–324. Malden, MA: Blackwell, 2002.

Rabinow, Paul. 'Representations Are Social Facts: Modernity and Postmodernity in Anthropology.' In James Clifford and George E. Marcus, eds, *Writing Culture: The Poetics and Politics of Ethnography*. 234–61. Berkeley: University of California Press, 1986.

Rainie, Lee. 'E-Citizen Planet.' Paper presented to the Personal Democracy Forum, Pace University, New York, 18 May 2007.

Rainie, Lee, and Mary Madden. 'Preliminary Findings from a Web Survey of Musicians and Songwriters.' Pew Internet & American Life Project, Pew Research Center, Washington, DC, May 2004.

Rakow, Lana F., and Laura A. Wackwitz. 'Voice in Feminist Communication Theory.' In Lana F. Rakow and Laura A. Wackwitz, eds, *Feminist Communication Theory: Selections in Context.* 93–110. London: Sage, 2004.

Rayfield, John R. 'What Is a Story?' *American Anthropologist* 74 (October 1972): 1085–106.

Renov, Michael. 'The End of Autobiography or New Beginnings?' In Jan Campbell and Janet Harbord, eds, *Temporalities, Autobiography and Everyday Life: Autobiography and Everyday Life.* 280–91. Manchester: Manchester University Press, 2002.

– *The Subject of Documentary.* Minneapolis: University of Minnesota Press, 2004.

RFC editor, et al. 'RFC 2555: 30 Years of RFCs.' The Internet Society, 7 April 1999.

Rich, Michael, Steven Lamola, Jason Gordon, and Richard Chalfen. 'Video Intervention/Prevention Assessment: A Patient-Centered Methodology for Understanding the Adolescent Illness Experience.' *Journal of Adolescent Health* 27 (2000): 155–65.

Richler, Noah. 'The Canadian Year in Video (Welcome to the Post-Fact Society).' *National Post*, 29 December 2008.

Rid, Thomas. 'The Bundeswehr's New Media Challenge.' *Military Review* (July-August 2007): 104–9.

Ridings, Catherine M., and David Gefen, 'Virtual Community Attraction: Why People Hang Out Online.' *Journal of Computer-Mediated Community* 10 (November 2004).

Rivlin, Gary. 'Hate Messages on Google Site Draw Concern.' *New York Times*, 7 February 2005.

Rizzo, Teresa. 'YouTube: The New Cinema of Attractions.' *Scan: Journal of Media Arts Culture* 5 (May 2008).

Roach, Catherine M. *Stripping, Sex, and Popular Culture.* London: Berg, 2007.

Roberts, Greg. 'Police Stat Analysis Claims Facebook, YouTube Fuelling Crime Wave.' *Herald Sun* (Victoria, Australia), 8 July 2008.

Rodriguez, Clemenca. *Fissures in the Mediascape: An International Study of Citizens' Media.* Cresskill, NJ: Hampton Press, 2001.

Rogers, Georgie. 'Piracy Still Prevailing.' BBC Radio, 16 January 2009.

Ronai, Carol R., Barbara A. Zsembik, and Joe R. Feagin. *Everyday Sexism in the Third Millennium.* London: Routledge, 1997.

Rowan, John, and Mick Cooper. *The Plural Self: Multiplicity in Everyday Life.* London: Sage, 1999.

Ruby, Jay. 'The Imagined Mirror: Reflexivity and the Documentary Film.' In
Alan Rosenthal and John Corner, *New Challenges for Documentary*. 2nd ed.
34–47. Manchester: Manchester University Press, 2005.

Ruffini, Patrick. 'The Internet is TV. Twitter is the Internet.' Personal Democracy Forum, techPresident.com, 18 December 2008.

Ruoff, Jeffrey. 'Home Movies of the Avant-Garde: Jonas Mekas and the New
York Art World.' In David James, ed., *To Free Cinema: Jonas Mekas and the New
York Underground*. 294–312. Princeton: Princeton University Press, 1992.

Salem, Arab, Edna Reid, and Hsinchun Chen. 'Content Analysis of Jihadi Ex-
tremist Groups' Videos.' In *Proceedings of the Intelligence and Security Informatics:
IEEE International Conference on Intelligence and Security Informatics*, San Diego,
CA, 23–24 May 2006.

San Miguel, Renay. 'YouTube Shines Spotlight on New High-Def Content.'
TechNewsWorld.com, 19 December 2008.

Saucier, Guylaine. 'The New CBC: A Commitment to Canadians.' Canadian
Broadcasting Corporation, 6 May 1999.

Schäfer, Mirko Tobias. 'Bastard Culture!: User Participation and the Extension
of Cultural Industries.' Ph.D. diss., Utrecht University, 2008.

Schejter, Amit M., and Moran Yemini. '"Justice, and Only Justice, You Shall Pursue": Network Neutrality, the First Amendment and John Rawls's Theory of
Justice.' *Michigan Telecommunications and Technology Law Review* 14 (December
2007): 137–74.

Scott, Anne, Lesley Semmens, and Lynette Willoughby. 'Women and the Inter-
net: The Natural History of a Research Project.' In Eileen Green and Alison
Adam, eds, *Virtual Gender: Technology, Consumption and Identity*, 3–27. London:
Routledge, 2001.

Scott, Ian. *American Politics in Hollywood Film*. London: Fitzroy Dearborn, 2000.

Seitz, Matt Zoller. 'Copy Rites: YouTube vs. Kevin B. Lee.' The House Next
Door Blog, 13 January 2009.

Sekula, Allan. *Photography Against the Grain: Essays and Photo Works, 1973–1983*.
Halifax: Nova Scotia College of Art and Design, 1984.

Seltzer, Wendy. 'DMCA "Repeat Infringers": Scientology Critic's Account Reinstated after Counter-Notification.' ChillingEffects.org, 6 June 2008.

Serfaty, Viviane. 'Online Diaries: Towards a Structural Approach.' *Journal of
American Studies* 38 (2004): 457–71.

Shade, Leslie Regan. 'Whose Global Knowledge?: Women Navigating the Net.'
Development 46 (March 2003): 49–54.

Shah, Chirag. 'YouTube Crawling: A VidArch Year in Retrospect.' The VidArch
Project, School of Information and Library Science, University of North
Carolina at Chapel Hill, 28 May 2008.

Shaidle, Kathy. 'Facebook Jihad.' *FrontPage Magazine*, 7 January 2009.

Shirky, Clay. *Here Comes Everybody: The Power of Organizing Without Organizations*. New York: Penguin Press, 2006.

Silverman, Toby. 'Expanding Community: The Internet and Relational Theory.' *Community, Work & Family* 4 (August 2001): 231–38.

Simon, Herbert A. 'Designing Organizations for an Information-Rich World.' in Martin Greenberger, ed., *Computers, Communication, and the Public Interest*. 39–72. Baltimore: Johns Hopkins University Press, 1971.

Singer, Ben. 'Early Home Cinema and the Edison Home Projecting Kinetoscope.' *Film History* 2 (1988): 37–69.

Sithigh, Daithi Mac. 'The Mass Age of Internet Law.' *Information and Communication Technology Law* 17 (June 2008): 79–94.

Slabbert, Nicholas J. 'Orwell's Ghost: How Teletechnology is Reshaping Civil Society.' *CommLaw Conspectus: Journal of Communications Law and Policy* 16 (2008): 349–59.

Smith, Aaron, and Lee Rainie. 'The Internet and the 2008 Election.' Pew Internet & American Life Project, Pew Research Center, Washington, DC, 15 June 2008.

Smith, Sidonie, and Julia Watson. *Interfaces: Women, Autobiography, Image, Performance*. Ann Arbor: University of Michigan Press, 2002.

Smith, Steve. 'Analysis: Mass Media, Magazine Influence Continue Declines.' Access Intelligence, minonline.com, 21 January 2009.

Snelson, Chareen. 'YouTube and Beyond: Integrating Web-Based Video into Online Education.' In *Proceedings of Society for Information Technology and Teacher Education International Conference*, Las Vegas, NV, 3 March 2008.

Solove, Daniel J. *The Future of Reputation: Gossip, Rumor, and Privacy on the Internet*. New Haven: Yale University Press, 2007.

Sparkes, Andrew C. 'Autoethnography: Self-Indulgence or Something More?' In Arthur P. Bochner and Carolyn Ellis, eds, *Ethnographically Speaking: Autoethnography, Literature, and Aesthetics*. 209–31. Lanham, MD: Altamira, 2009.

Spender, Dale. *Nattering on the Net: Women, Power and Cyberspace*. Toronto: Garamond Press, 1995.

Spicer, Kate, and Abul Taherreport. 'Girls and Young Women are Now the Most Prolific Web Users.' *Sunday Times* (London), 9 March 2008.

'Star Wars Kid is Top Viral Video.' BBC News, 27 November 2006.

Stauffer, Donald. *The Art of English Biography before 1700*. Cambridge, MA: Harvard University Press, 1930.

Steele, Francesca. 'The Web Watcher: Tricia Walsh-Smith Does It Again.' *Times* (London), 8 December 2008.

Stelter, Brian. 'Hulu Questions Count of Its Audience.' *New York Times*, 14 May 2009.

– 'MySpace Might Have Friends, but It Wants Ad Money.' *New York Times*, 16 June 2008.
– 'Web Suicide Viewed Live and Reaction Spur a Debate.' *New York Times*, 24 November 2008.
– 'YouTube Videos Pull In Real Money.' *New York Times*, 11 December 2008.
Stelter, Brian, and Brad Stone. 'Digital Pirates Winning Battle With Studios.' *New York Times*, 5 February 2005.
Stephens, Mitchell. *The Rise of the Image: The Fall of the Word.* Oxford: Oxford University Press, 1998.
Sterling, Greg. 'YouTube Video and Usage Facts.' *Sterling Marketing Intelligence*, 31 August 2006.
Stone, Brad. 'Accuser Says Web Site for Teenagers Has X-Rated Link.' *New York Times*, 11 July 2007.
Stone, Brad, and Ashlee Vance. '$200 Laptops Break a Business Model.' *New York Times*, 26 January 2009.
Strangelove, Michael. 'Redefining the Limits to Thought within Media Culture: Collective Memory, Cyberspace, and the Subversion of Mass Media.' Ph.D. diss., University of Ottawa, 1998.
– *The Empire of Mind: Digital Piracy and the Anti-capitalist Movement.* Toronto: University of Toronto Press, 2005.
– 'Virtual Video Ethnography: Towards a New Field of Internet Cultural Studies.' *Interin* 3 (June 2007).
'Survey Telecoms: A World of Connections.' *Economist*, 28 April 2007.
Szabo, Liz. 'Report: TV, Internet Harm Kids.' *USA Today*, 2 December 2008.
Taft, Michael. 'Men in Women's Clothes: Theatrical Transvestites on the Canadian Prairie.' In Pauline Greenhill and Diane Tye, eds, *Undisciplined Women: Tradition and Culture in Canada*. 131–50. Montreal: McGill-Queen's University Press, 1997.
Tanaka, Wendy. 'Gabbing with Top Googler.' *Forbes*, 11 June 2008.
Tannen, Deborah. *The Argument Culture: Stopping America's War of Words.* New York: Ballantine Books, 1999.
Tartakoff, Josephus. 'Analyst: YouTube Will Lose Almost $500 Million This Year.' *Washington Post*, 3 April 2009.
Terranova, Tiziana. 'Free Labor: Producing Culture for the Digital Economy.' *Social Text* 18, no. 2 (2000): 33–57.
'The First YouTube Election.' *Los Angeles Times*, 6 September 2006.
Thompson, John B. *The Media and Modernity: A Social Theory of the Media.* Stanford: Stanford University Press, 1995.
Tice, Diane M. 'Self-concept Change and Self-presentation: The Looking Glass Self Is Also a Magnifying Glass.' *Journal of Personality and Social Psychology* 63 (1992): 435–51.

Time. 13 December 2006.

Travers, Ann. 'Parallel Subaltern Feminist Counterpublics in Cyberspace.' *Sociological Perspectives* 46 (Summer 2003): 223–37.

Trujillo, Iván. 'La Filmoteca de la Universidad Nacional Autónoma de México.' In Karen L. Ishizuka and Patricia R. Zimmerman, eds, *Mining the Home Movie: Excavations in Histories and Memories*. 57–61. Berkeley: University of California Press, 2008.

Tryon, Chuck. '"Why 2008 Won't Be like 1984": Viral Videos and Presidential Politics.' Flowtv.org, 21 March 2007.

Turkheimer, Margot. 'A YouTube Moment in Politics: An Analysis of the First Three Months of the 2008 Presidential Election.' Urban and Environmental Policy Institute, Occidental College, 2007.

Turnbull, Giles. 'The Seven-Year-Old Bloggers.' BBC News, 14 June 2004.

Ulanoff, Lance. 'Lose the Optical Drives in Laptops.' *PC Magazine*, 28 May 2008.

van Dijck, José. 'Television 2.0: YouTube and the Emergence of Homecasting.' Paper presented to the Creativity, Ownership and Collaboration in the Digital Age, Cambridge, Massachusetts Institute of Technology, 27–9 April 2007.

– 'Users Like You? Theorizing Agency in User-generated Content.' *Media, Culture & Society* 31 (January 2009): 41–58.

Van Dijk, Teun A. *Discourse and Structures of Power*. New York: Macmillan, 2008.

Van Zoonen, Liesbet. 'Feminist Internet Studies.' *Feminist Media Studies* 1 (March 2001): 67–72.

Waldfogel, Joel. 'Lost in the Web: Does Web Distribution Stimulate or Depress Television Viewing?' NBER Working Paper No. W13497, October 2007.

Walker, Jill. 'Mirrors and Shadows: The Digital Aestheticisation of Oneself.' In *Proceedings, Digital Arts and Culture 2005*, IT University, Copenhagen, December 2005.

Walkerdine, Valerie. 'Femininity as Performance.' In Lynda Stone and Gail Masuchika Boldt, eds, *The Education Feminism Reader: Developments in a Field of Study*. 57–72. London: Routledge, 1994.

Wallace, Patricia. *The Psychology of the Internet*. Cambridge: Cambridge University Press, 1999.

Wallsten, Kevin. '"Yes We Can": How Online Viewership, Blog Discussion and Mainstream Media Coverage Produced a Viral Video Phenomenon.' Paper presented to the Annual Meeting of the American Political Science Association, Boston, 2008.

Walters, Ben. 'The Online Stump.' *Film Quarterly* 61 (October 2007): 60–1.

Warmbrodt, John, Hong Sheng, and Richard Hall. 'Social Network Analysis of

Video Bloggers' Community.' *Proceedings of the 41st Annual Hawaii International Conference on System Sciences*, Waikoloa, HI, January 2008.

Wartella, Ellen, and Michael Robb. 'Historical and Recurring Concerns about Children's Use of the Mass Media.' In Sandra L. Clavert, ed., *The Handbook of Children, Media, and Development*. 7–26. London: Blackwell, 2008.

Watson, Nessim. 'Why We Argue About Virtual Community: A Case Study of the Phish.Net Fan Community.' In Steve Jones, ed., *Virtual Culture: Identity and Communication in Cybersociety*. 102–32. London: Sage, 1997.

Waxman, Sharon. 'Hollywood 2.0 with Barry Diller.' TheWrap.com, 29 January 2009.

Wayne, Mike. *Theorising Video Practice*. London: Lawrence & Wishhart, 1997.

Weissman, Kristin Noelle. *Barbie: The Icon, the Image, the Ideal: An Analytical Interpretation of the Barbie Doll in Popular Culture*. Boca Raton, FL: Universal, 1999.

Wesch, Michael. 'The State of Research.' Presentation at the 24/7 DIY Video Summit Conference, University of Southern California, February 2007.

Westin, Alan F. *Privacy and Freedom*. New York: Atheneum, 1967.

Wharton, Rachel. 'A Click of the Mousse.' *NY Daily News*, 27 July 2007.

White, Mimi. *Tele-Advising: Therapeutic Discourse in American Television*. Chapel Hill: University of North Carolina Press, 1992.

Whittaker, Steve, Ellen Issacs, and Vicki O'Day. 'Widening the Net. Workshop Report on the Theory and Practice of Physical and Network Communities.' *SIGCHI Bulletin* 29 (July 1997): 27–30.

Wiener, Wendy J., and George C. Rosenwald. 'A Moment's Monument: The Psychology of Keeping a Diary.' In Ruthellen Josselson and Amia Lieblich, eds, *The Narrative Study of Lives*. 30–58. London: Sage, 1993.

Williams, Linda. *Hard Core: Power, Pleasure, and the 'Frenzy of the Visible.'* Berkeley: University of California Press, 1989.

Williams, Raymond. *Television, Technology and Cultural Form*. London: Fontana, 1974.

Wilson, Samuel M., and Leighton C. Peterson. 'The Anthropology of Online Communities.' *Annual Review of Anthropology* 31 (June 2002): 449–67.

Wolf, Eric R. *Pathways of Power: Building an Anthropology of the Modern World*. Berkeley: University of California Press, 2001.

Wood, Peter. *A Bee in the Mouth: Anger in America Now*. New York: Encounter Books, 2006.

Worth, Sol. 'Margaret Mead and the Shift from "Visual Anthropology" to "the Anthropology of Visual Communication."' *Studies in Visual Communication* 6 (Spring 1980): 15–22.

– 'Toward an Anthropological Politics of Symbolic Forms.' In Dell Hymes, ed., *Reinventing Anthropology*. 335–64. New York: Pantheon, 1973.

Wortham, Jenna. 'Cheating Fans Give Avril Lavigne a YouTube Lift.' *Wired*, 24 June 2008.

Worthen, Ben. 'Cisco Says Internet Video to Explode.' *Wall Street Journal*, 9 June 2009.

Wright, Karen. 'Dare to Be Yourself.' *Psychology Today*, May 2008.

Wright, Michelle M. *Becoming Black: Creating Identity in the African Diaspora*. Durham, NC: Duke University Press, 2004.

Wylie, Margie. 'No Place for Women: Internet is a Flawed Model of the Infobahn.' *Digital Media* 4 (January 1995): 3–7.

Yeung, King-To, and John Levi Martin. 'The Looking Glass Self: An Empirical Test and Elaboration.' *Social Forces* 81 (March 2003): 843–79.

Young, Jeffrey R. 'An Anthropologist Explores the Culture of Video Blogging.' *Chronicle of Higher Education* 53 (May 2007).

YouTube. 'Terms of Use.' youtube.com, 2008.

'YouTube Contributes to the War on Terror.' *Layalina Review on Public Diplomacy and Arab Media* 4 (September 2008).

'YouTube is …' YouTube.com, Audience One-Sheeter, 2008.

YouTube Team. 'A YouTube for All of Us.' Broadcasting Ourselves: The YouTube Blog, 2 December 2008.

Zalis, Elayne. 'At Home in Cyberspace: Staging Autobiographical Scenes.' *Biography* 26 (Winter 2003): 84–119.

Zimmerman, Patricia R. 'Geographies of Desire: Cartographies of Gender, Race, Nation and Empire in Amateur Film.' *Film History* 8 (1996): 85–98.

– 'Hollywood, Home Movies, and Common Sense: Amateur Film as Aesthetic Dissemination and Social Control, 1950–1962.' *Cinema Journal* 27 (Summer 1988), 23–44.

– 'The Home Movie Movement: Excavations, Artifacts, Minings.' In Karen L. Ishizuka and Patricia R. Zimmerman, eds, *Mining the Home Movie: Excavations in Histories and Memories*. 1–28. Berkeley: University of California Press, 2008.

– *Reel Families: A Social History of Amateur Film*. Bloomington: Indiana University Press, 1995.

Zoglin, Richard and Georgia Harbison. 'Goodbye to the Mass Audience.' *Time*, 19 November 1990.

Zuern, John. 'Online Lives: Introduction.' *Biography* 26 (Winter 2003): v–xxv.

Index of Names

Index of Subjects

ABC, 183

Abu Ghraib, 153, 154

activism: fat, 94, 95; media, 43; political, 178; YouTube genre of, 125

Adobe, 185

adolescents. *See* teenagers

Adventures of Ozzie and Harriet, The, 49

advertising, 5–6, 50, 72, 74, 159, 175, 178, 203n41; and amateur videographers, 7, 134, 174, 177, 182; and children, 54; and femininity, 94; Google, 6; and MySpace, 195n7; online, 164, 169; political, 140; techniques of, 74; television, 167–8. *See also* product placement; YouTube: advertising

aesthetics: of amateur video, 30, 42; of attraction, 47; avant-garde, 43; of commercial studios, 28, 36, 146, 181; of hierarchies of, 84; home movies, 27, 28, 32, 40; intolerance, 37, 40; mainstream, 87; of negation, 174; norms 39; of transparency, 70; of YouTube, 26;

AI Lab Dark Web, 151

alcohol, 62

Al Jazeera, 153

amateur film, 24; definition of, 23, 197n3; early, 127. *See also* home movies

amateur video. *See* video, amateur online

amateurs: and appropriation, 162, 182; definition of, 3, 16–17, 177; and Google, 7; as irrelevant, 183; media practice of, 15, 43, 44, 73, 138; motives, 4, 23; as producers, 14, 17, 30, 173; vs professional, 82, 113; and reality, 180; significance of, 174–5; and YouTube, 16, 113, 121, 136, 157. *See also* cultural production: amateur

America, 41, 68, 88, 146; anti-, 129; broadband use, 127; cultural ideal of, 119; elections, 138; and hatred, 147; and homemade bombs, 151; Internet population, 146; lifestyle 134–5; public discourse, 119

America Online, 184

analogue home movies. *See* home movies

anger, 119. *See also* hate

anorexia. *See* pro-anorexia

Index of YouTube Videos

Digital Futures is a series of critical examinations of technological development and the transformation of contemporary society by technology. The concerns of the series are framed by the broader traditions of literature, humanities, politics, and the arts. Focusing on the ethical, political, and cultural implications of emergent technologies, the series looks at the future of technology through the 'digital eye' of the writer, new media artist, political theorist, social thinker, cultural historian, and humanities scholar. The series invites contributions to understanding the political and cultural context of contemporary technology and encourages ongoing creative conversations on the destiny of the wired world in all of its utopian promise and real perils.

Series Editors:
Arthur Kroker and Marilouise Kroker

Editorial Advisory Board:
Taiaiake Alfred, University of Victoria
Michael Dartnell, University of New Brunswick
Ronald Deibert, University of Toronto
Christopher Dewdney, York University
Sara Diamond, Banff Centre for the Arts
Sue Golding (Johnny de philo), University of Greenwich
Pierre Levy, University of Ottawa
Warren Magnusson, University of Victoria
Lev Manovich, University of California, San Diego
Marcos Novak, University of California, Los Angeles
John O'Neill, York University
Stephen Pfohl, Boston College
Avital Ronell, New York University
Brian Singer, York University
Sandy Stone, University of Texas, Austin
Andrew Wernick, Trent University

Books in the Series:
Arthur Kroker, *The Will to Technology and the Culture of Nihilism: Heidegger, Nietzsche, and Marx*
Neil Gerlach, *The Genetic Imaginary: DNA in the Canadian Criminal Justice System*
Michael Strangelove, *The Empire of Mind: Digital Piracy and the Anti-Capitalist Movement*
Tim Blackmore, *War X: Human Extensions in Battlespace*
Michael Dartnell, *Insurgency Online: Web Activism and Global Conflict*
Janine Marchessault and Susan Lord, eds., *Fluid Screens, Expanded Cinema*
Barbara Crow, Michael Longford, and Kim Sawchuk, eds., *The Wireless Spectrum: The Politics, Practices, and Poetics of Mobile Media*
Michael Strangelove, *Watching YouTube: Extraordinary Videos by Ordinary People*